Planning, Housing and Infrastructure for Smart Villages

Some 7.3 billion people currently live on the planet. Of these, 3.4 billion live in rural areas. In just a few regions—Latin America, the Middle East and North Africa—less than 50 per cent of poverty is now located in rural areas. But for the rest of the world's regions between 55 per cent and 80 per cent of the poor continue to live in the countryside. Progress is being made, but much of the knowhow needed is not disseminated outside of a small coterie of professionals who work in the area. With urban development attracting a great deal of attention lately, poorer rural areas deserve the same and new knowledge for empowerment of rural communities is urgently needed.

This book provides an overview of current thinking and practices that have emerged over the last thirty years for uplifting rural communities in developing economies. Drawing on a body of knowledge across a spectrum of relevant disciplines, this book provides a range of innovative ideas for rural planning, housing and infrastructure development.

Governments in many emerging economies, where rural poverty is often most acute, have attempted to improve livelihoods. Approaches and techniques that have been used for urban development are often not applicable to rural communities. Studies show that money allocated for rural development is often not effectively spent due to distance, lack of infrastructure, lack of education, poverty and other factors. Meanwhile, the gap in development between the city and country continues to grow, sometimes leading to social and political instability, in both developing and developed countries. This book seeks to provide a guidebook for meeting such challenges.

Through in-depth enquiry of global practices and thinking about rural development, and selected case studies, the authors argue that careful consideration must be given to incorporating issues of resilience, resourcefulness and the involvement of communities at grassroots levels in realising the transformation of rural settlements into Smart Villages.

Hemanta Doloi is a senior lecturer in the Faculty of Architecture, Building and Planning at the University of Melbourne, Australia. Dr Doloi is the founding director of the Smart Villages Lab (SVL) and the lead investigator of the Smart Villages Project, sponsored by the Government of Assam, India, for developing solutions on affordable housing and infrastructure systems.

Ray Green is a professor in the Faculty of Architecture, Building and Planning at the University of Melbourne, Australia. He is the author of *Coastal Towns in Transition* (2010) and co-author of *The Green City: Sustainable Homes, Sustainable Suburbs* (2005). His research has been published in urban planning, landscape architecture and environmental psychology journals.

Sally Donovan is a research fellow with the Smart Villages Lab in the Faculty of Architecture, Building and Planning at the University of Melbourne, Australia. Dr Donovan has over ten years' experience researching environmental management and environmental policy development.

Planning, Housing and Infrastructure for Smart Villages

Hemanta Doloi, Ray Green and Sally Donovan

Routledge
Taylor & Francis Group
LONDON AND NEW YORK

First published 2019
by Routledge
2 Park Square, Milton Park, Abingdon, Oxon OX14 4RN

and by Routledge
52 Vanderbilt Avenue, New York, NY 10017

Routledge is an imprint of the Taylor & Francis Group, an informa business

British Library Cataloguing-in-Publication Data
A catalogue record for this book is available from the British Library

Library of Congress Cataloging-in-Publication Data
Names: Doloi, Hemanta Kumar, author. | Green, Ray, (Professor of Architecture) author. | Donovan, Sally, 1953- author.
Title: Planning, housing and infrastructure for smart villages / Hemanta Doloi, Ray Green and Sally Donovan.
Description: Abingdon, Oxon ; New York, NY : Routledge is an imprint of the Taylor & Francis Group, an Informa Business, 2019. | Includes bibliographical references.
Identifiers: LCCN 2018035682| ISBN 9780815365648 (hbk) | ISBN 9780815365655 (pbk) | ISBN 9781351261081 (ebk)
Subjects: LCSH: Smart cities. | City planning. | Housing. | Infrastructure (Economics)
Classification: LCC TD159.4 .D65 2019 | DDC 307.72—dc23
LC record available at https://lccn.loc.gov/2018035682

ISBN: 978-0-8153-6564-8 (hbk)
ISBN: 978-0-8153-6565-5 (pbk)
ISBN: 978-1-351-26108-1 (ebk)

Typeset in Times New Roman
by Swales & Willis Ltd, Exeter, Devon, UK

Contents

Figures

Tables

Preface

Of the 7.3 billion people currently living on the planet, 3.4 billion of them live in rural areas. A large portion of these people, somewhere between 55 per cent and 80 per cent depending upon where, are poorer than their urban counterparts. Progress is being made in many places to bridge this gap, however, much of the know-how needed to improve the living conditions of rural communities has been lacking, or is not being optimally utilised. Ways for improving the conditions of the urban poor has attracted a great deal of research attention in recent times. However, research focused on improving living conditions for people living in poorer rural areas has not received as much attention. Yet there is an urgent need for this type of research to inform the development of policies and governance structures that can most effectively deliver improvements for rural communities. Over the past couple of decades, in many parts of the developing world, grand schemes targeting improvements in rural development have either failed or simply have not delivered the expected outcomes. At the same time, the gap between the quality of life of people in rural communities and urban populations has grown wider.

This book provides an overview of the many interrelated factors that need to be considered in creating Smart Villages, the aim of which is to improve living conditions for people living in rural settlements, primarily in developing countries. It provides an introduction into various considerations and approaches for income generation, education, agriculture, water, sanitation, housing, energy, waste, health and governance, all of which need to be considered in achieving this goal. The book provides an overview of the intricacies of each of these factors and sets a framework for further explorations. It also proposes that new technologies be blended with traditional forms of rural development with the overarching objective to make rural villages better places to live and as ecologically, socially and economically sustainable as possible.

This book was initially inspired by research funded by the State Government of Assam, India, which is focusing on "smart" approaches to the planning and design of rural settlements in that state. The book is authored by three researchers based in the Faculty of Architecture Building and Planning at the University of Melbourne who are working on that project in collaboration with researchers at the Assam Engineering College and Indian Institute of Technology Guwahati, both located in Assam, India. The aim of that project, and the Smart Villages Lab (SVL) in which it is being undertaken, is to promote improvements in living standards for rural inhabitants in the state through the incorporation of best practice approaches to planning, housing and infrastructure development of rural settlements.

Acknowledgements

The authors wish to express their profound gratitude and sincere appreciation to the Government of Assam for sponsoring the Smart Villages research at the University of Melbourne and providing opportunities for creating a shared capacity in rural development, an area of critical need. The authors are particularly grateful to the Ex Chief Minister of Assam, Mr Tarun Gogoi, for his vision and personal interest in supporting the research and collaboration with the University of Melbourne. Being the key academic partner in this effort, the technical input and administrative support offered by Dr Atul Bora, Principal of Assam Engineering College and past Director of Technical Education (in-charge), is highly commendable. Support extended by the senior officials of the State of Assam, namely Shri V. K. Pipersenia, Shri Pabon Borthakur, Shri MGK Bhanu, Mr Ranjit Hazarika, Mr Ravi Kota, Shri Rajiv Bora, Shri Manish Thakur and many others in materialising the project from concept to reality, is much appreciated and will never be forgotten.

The authors particularly extend their thanks and appreciation to the Honourable Minister of Education, Dr Himanta Biswa Sarma, and the Honourable Chief Minister, Shri Sarbananda Sonowal, for their unconditional support for this project from the first day in the office. The vision and mission of these visionary leaders is making a real difference in rural communities in the state of Assam through their support of scientific research in partnership with The University of Melbourne and its Smart Village Lab. The current Principal Secretary Education, Mr Ajay Tiwari, Principal Secretary Panchayat and Rural Development, Dr John Ekka, and Director of Technical Education (in-charge), Ms Krishna Gohain deserve special praise for extending their continuous support and guidance in helping to steer the project. The authors also sincerely appreciate the present Minister of Education, Mr Siddhartha Bhattacharya, for his inspiration and support in the Smart Villages Project.

A special thank you goes to David Week for his involvement in the development of the book proposal and Prof. Mark Burry, past professor of Urban Future at the University of Melbourne, for his kind assistance and guidance given to the project leader, Dr Hemanta Doloi in the early stage of the Smart Villages Project. The authors also highly appreciate two distinguished colleagues, Associate Professor Robert Crawford and Dr Dominique Hes in the Faculty of Architecture Building and Planning at the University of Melbourne for their involvement and support in the project. Prof. Daryl LeGrew, past interim Dean, and Prof. Paul Walker, Associate Dean Research at the Faculty of Architecture, Building and Planning, deserve special thanks for their kind support and guidance in helping to set up the project. The leadership and facilitation of the current Dean, Prof. Julie Willis, Deputy Dean, Associate Professor Andrew Hutson, and the Faculty's Executive Director, Rebecca Bond, is particularly commendable.

Finally, the authors express their sincere gratitude to many experts who shared valuable experience and provided numerous examples of rural development initiatives that have enhanced the content of this book. In particular, they would like to thank Dr Kiran Shinde for his input. The publishers also deserve special thanks for their kind support and continued interest in the project.

1 Introduction

Rural settlements

Rural settlements, which are home to more than 40 per cent of the world's population, are a lot of things – repositories of culture, producers of food for urban populations and in some cases stewards of the natural environment. *Planning, Housing and Infrastructure for Smart Villages* seeks to shed light on the inter-related aspects of rural development in the context of developing countries. The Smart Village Model (SVM), as presented in this book, envisions an integrated approach to the design of rural settlements that aims to achieve improvements in living standards for inhabitants while simultaneously making their homes, villages and lifestyles more environmentally, socially and economically sustainable in the face of ever-increasing environmental and societal changes. The SVM presented here combines the best attributes of traditional forms of rural development and design with newer, "smart" approaches. The word "smart", as conjured by notions of "smart cities" and "smart homes", has become synonymous with internet connectivity and digital technology – smartphones and other digital devices, smart TVs, and household appliances you can interact with remotely. Digital technology and internet connectivity have a key role to play in helping rural communities adapt to the 21st century, however, the SVM presented here promotes an approach to the design of rural settlements that goes beyond simple adoption of digital technology and proposes a range of initiatives aimed at providing a better quality of life for rural communities, one that allows people to thrive and not simply survive. It is not a "one-size-fits-all" model, but seeks solutions to problems tailored to local environments, people and the specific problems each community is facing.

What is a Smart Village?

A Smart Village is one where people's housing and infrastructure needs are met, resulting in improved comfort and functionality for residents yet still being affordable. It advocates for houses that take advantage of new technologies while still reflecting traditional ways of design, construction and operation where this is appropriate.

Cultural sustainability is an important dimension of the Smart Village Model. Many rural communities, often more remote communities, have been able to hold on to their cultural traditions. In urban areas such traditional culture has often been lost or distorted for commercial gain. Preserving and/or reviving cultural traditions is one way a village can achieve economic sustainability, for example through promotion of cultural and ecotourism or the production and sale of crafts and other items derived from local resources. Many rural settlements have, however, lost, or are in the process of losing, connections with their traditional cultures due to the pervasive forces of globalisation. Therefore, reviving and conserving a village's tangible and intangible heritage, in the context of the homogenising effects of ever-expanding urban economies and cities, is a key factor in the making of a Smart Village.

Smart Villages are also places that are economically sustainable, made possible, for example, through providing educational opportunities relevant to the activities that are essential for taking advantage of economic opportunities.

Resilience to withstand disruptions, and preparing for future changes, is another aspect of a Smart Village. Future changes may not be like anything seen in historical times and rural communities need to be prepared. In this respect, the notion of a Smart Village is concerned with working in harmony with the natural environment, understanding the risks it can present, conserving nature, reducing greenhouse gases, wisely managing waste, and many other actions that can make a village more environmentally sustainable.

Many rural communities are experiencing decreasing and aging populations, and this is limiting their ability to maintain sustainable workforces over time. Mitigating this outward migration of people from rural to urban areas is key to the economic survival and sustainability of many villages. But stopping this outward migration requires offering people, particularly younger people, reasons why they would want to stay and live in their village rather than seeking employment in a city that is often far removed from their rural homes. This outward migration also poses a significant threat to the sense of community and social connectivity within villages, another integral aspect of any Smart Village.

Rural people who work within the limits of the finite resources available to them, and still find ways to provide for themselves and their families on a day to day basis, possess knowledge passed down through generations, sometimes over thousands of years, that can be tapped into in creating Smart Villages. The idea is to find innovative solutions to problems facing rural communities that blend traditional and new technology in the planning, design and management of villages for the future.

Matthew Harris, in the School of Public Health, Imperial College London (Bhatti et al., 2017), suggests: "There appears to be an unspoken attitude that poorer countries learn from richer ones and not the other way around." During his research into healthcare practices in developing economies he discovered a range of innovative solutions for medical procedures based on local knowledge, ingenuity and available resources. For example, surgeons in the United Kingdom use orthopaedic drills during surgery, which are made with heat resistant materials that allow them to be easily sterilised between patients, but at a cost of around 30,000 Pounds Sterling.

In Uganda, where money is tighter, surgeons use ordinary household drills, wrapped in a sterilisable material, to achieve the same result for much less money. Such simple innovations are often highly effective and more affordable solutions. Simple innovations eventually result in lesser reliance on assistance from outside bodies such as governments and NGOs.

Millennium Development Goals

The Millennium Development Goals (United Nations, 2015a) are laid out in a document signed by the leaders of 191 nations, who committed to working towards achieving them by 2015. The eight key goals outlined in this document directly relate to objectives of creating Smart Villages and include:

1 Eradicating extreme poverty and hunger

- The number of people living on less than 1.25 USD a day has been reduced from 1.9 billion in 1990 to 836 million in 2015, narrowly missing the target of halving the proportion of people suffering from hunger worldwide.

2 Achieving universal primary education

- Primary school enrolment figures have increased significantly from 83 per cent in 2000 to 91 per cent in 2015, but not quite reaching the goal.

3 Promoting gender equality and empowerment of women

- Around two thirds of developing countries achieved gender parity in primary education.

4 Reducing child mortality

- Child mortality fell from 90 to 43 deaths per 1000 births, not quite reaching the two thirds target.

5 Improving maternal health

- This also fell by around a half, not quite reaching the goal of two thirds.

6 Combating HIV/AIDs, malaria, and other preventable diseases

- HIV/Aids infection fell by approximately 40 per cent between 2000 and 2013.

7 Ensuring greater environmental sustainability

- The target of halving the proportion of people without access to potable water was achieved in 2010, five years ahead of schedule, but 663 million people still do not have access to clean and readily available water.

8 Developing global partnerships to achieve better outcomes

- Overseas development assistance to developing countries has increased by 66 per cent from 2000 to 2014.

Although the world has come a long way towards achieving the eight goals outlined above, distribution has been uneven, and inequalities between urban and rural settlements and their communities exist. A new set of 17 Sustainable Development Goals came out in 2015 (United Nations, 2015c). These incorporate the Millennium Development Goals plus include more targeted goals specific to environmental problems such as greenhouse gas emissions and waste management. The aim is for the signature countries to have achieved these goals by 2030. The new targets are more concerned with reaching poorer rural communities, rather than the "low hanging fruit" of the urban voting population in densely populated urban areas as opposed to rural and remote areas that represent fewer votes.

Discrepancies between rural and urban areas can sometimes be dramatic. For example, only 56 per cent of births in rural areas are attended to by skilled health personnel compared to 87 per cent in urban areas (United Nations, 2015b). About 16 per cent of people in rural areas do not have access to sources of safe drinking water, compared with only 4 per cent of urban populations. Some 50 per cent of people living in rural areas lack improved sanitation facilities compared to only 18 per cent of people in urban areas. The average out-of-school rate in rural areas, based on a survey of 63 developing countries from 2008 to 2012, was twice as high (16 per cent) as it is in urban areas (8 per cent). Mortality is also higher for children in rural areas compared to urban areas. In fact, rural children are some 1.7 times more likely to die before their fifth birthday than children in urban areas.

Maternal education remains the most powerful determinant of inequality in terms of survival of children in rural as opposed to urban areas. Mothers with a secondary or higher education are almost three times more likely to have children that survive past five years than do mothers with no education. In addition, there is a disparity between rural and urban in knowledge about HIV and its transmission among communities (23 per cent to 36 per cent).

Notably, rural areas have, however, been changing rapidly across numerous fronts and improving community bottom lines. For example, the proportion of the global rural population without access to improved drinking water has declined by more than half since 1990, from 38 per cent to 16 per cent in 2015. Since 1990 the proportion of the global rural population without access to improved sanitation has declined by nearly a quarter, and open defecation rates in rural areas have fallen from 38 per cent to 25 per cent by 2015. Still, nearly half of the people living in rural areas do not have access to improved sanitation facilities, and one in four still practise open defecation.

Globally, the proportion of people owning a mobile phone was 97 per cent in urban areas by 2015, but only 64 per cent in rural areas with an estimated 450 million people still living in rural areas with no internet connectivity.

Making a Smart Village

A Smart Village is one that meets not only the UN development goals as outlined above but also many other relevant location-specific goals found in the Organisation for Economic Co-operation and Development (OECD) regional well-being framework (Organisation for Economic Co-operation and Development, 2016). The first step is identifying the current status of a village: what are the current problems they are facing and what steps do they need to take to be improved? Are there barriers that could be limiting the success of implementing the types of steps called for in addressing problems? If so, how can we avoid/overcome these barriers?

In summary, a Smart Village is one that has adequate reference to all the key elements as depicted in Figure 1.1:

- Jobs/Income: A Smart Village is one where everyone has an income sufficient to maintain a healthy lifestyle.
- Education: In a Smart Village everyone is literate and has access to learning opportunities, which can be delivered face-to-face or through the internet, and which allow skills to be acquired to gain employment.
- Housing: A Smart Village is one where everyone has somewhere to live, which is warm and dry, clean and safe.
- Healthcare: A Smart Village has adequate sanitation, with clean water available for drinking and washing and where waste is treated in a safe and sustainable manner. It is a place where medical care is available to everybody on both an emergency and regular care basis.
- Environment: A Smart Village is one that is in harmony with the natural environment, and a place where natural environments are conserved, and degraded environments are restored wherever possible. This means providing appropriate and safe habitats for wildlife and preserving the ecological patterns that shape those habitats.
- Access to services: A Smart Village has enough energy, ideally from carbon neutral sources, for everybody's cooking, heating and refrigeration needs. This also applies to healthcare and educational facilities in terms of safe food handling and management and for cooling and heating. Waste removal, recycling and treatment (e.g. as a potential source of employment, fertiliser and an energy source) is implemented wisely. Communication services (mail, telephone, television, radio), and importantly the internet, are integrated into the daily lives of people, for example to inform them of potential dangers (incoming bad weather etc.) and access to neighbours, relatives and government bodies.
- Security and safety: A Smart Village is safe and resilient to natural and human threats, both seen (e.g. deforestation) and unseen (e.g. ground water pollution).
- Civic engagement: A Smart Village is one in which the citizens are engaged in the daily function of their village and surrounding landscape.

The following chapters focus on the different factors in the creation of a Smart Village as outlined above. Chapter 1 has introduced the notion of a Smart Village and its various dimensions. Chapter 2 is concerned with income generation, with an emphasis on ways of generating income that are socially, environmentally and

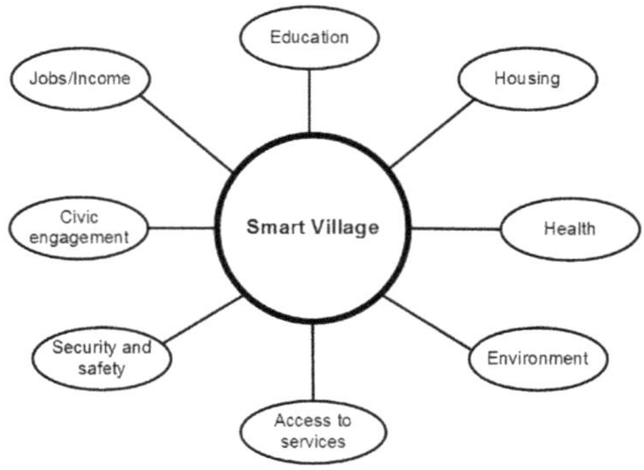

Figure 1.1 Idealised model of a Smart Village

economically sustainable. One means of income generation that can be suitable for some rural communities, particularly those located in natural environmental settings, is ecotourism, which can also help to reinforce the sustainable construction and operation of villages. Education, essential to providing the knowledge and skills necessary to help people break out of the poverty cycle, is discussed in Chapter 3. Chapter 4 discusses agriculture, a primary means of income generation for many rural commodities, and how it can be undertaken in a sustainable manner. Chapter 5 is concerned with housing, with an emphasis on affordable and sustainable forms of housing that provide people with "homes" and that respect the wisdom embodied in vernacular forms of architecture. Chapter 6 discusses energy and its sustainable generation and use. Access to energy is considered by the United Nations as a basic human right yet many rural communities still do not have access to electricity and often rely solely on burning fuels, such as wood and dung, for their cooking and heating needs. This chapter also investigates various ways of providing affordable, alternative sources of energy to remote communities. Chapter 7 is concerned with waste management, which in most rural communities is largely organic in nature, which if dumped untreated into the environment can damage both the environment and pose serious risks to human health. Yet organic waste also contains valuable nutrients that are beneficial to soils, an important consideration in sustainable agriculture. Organic waste can also potentially be used for energy generation. Water and sanitation, which are often linked, are discussed in Chapter 8, both of which are important as a lack of access to clean drinking water and poor hygiene are leading causes of disease, lifelong disabilities and premature death. Affordable and sustainable options of providing access to clean water and sanitation facilities are, therefore, vitally important to smart rural development. Chapter 9 addresses transportation, with an emphasis on how a lack of good transportation infrastructure can severely inhibit the ability of rural communities to generate income. It argues that rural communities should, where possible, take responsibility for maintaining their own transportation infrastructure, thus helping to ensure its long-term viability. Health and well-being are addressed in Chapter 10. Many diseases that are leading causes of disability and death in rural areas, have, in many cases, largely been eradicated in more affluent regions of the world. Chapter 11 looks at the role of the natural environment in rural developments, whose inhabitants typically live in close proximity to precious natural resources they need for survival. This chapter explores ways rural communities can work to conserve, and where possible enhance, the natural environment. Finally, Chapter 12 explores governance structures that can effectively implement the strategies discussed in previous chapters and which are required to bring the Smart Village Model to life.

References

Bhatti, Y., Taylor, A., Harris, M., Wadge, H., Escobar, E., Prime, M., Patel, H., Carter, A. W., Parston, G., Darzi, A. W. & Udayakumar, K. 2017. Global lessons in frugal innovation to improve health care delivery in the United States. *Global Health Policy*, 36, 1912–1919.

Organisation for Economic Co-operation and Development 2016. OECD regional well-being: A user's guide. *In*: OECD (ed.). Paris, France.

United Nations 2015a. *Millennium development goals and beyond 2015* [Online]. United Nations. Available: www.un.org/millenniumgoals/bkgd.shtml [Accessed 21 December 2017].

United Nations 2015b. The millennium development goals report. New York.

United Nations 2015c. *Sustainable development goals* [Online]. United Nations. Available: www.un.org/sustainabledevelopment/ [Accessed 1 February 2018].

2 Income generation

Introduction

Lack of income generating opportunities is at the heart of the urban-rural divide. It is driving young people, especially men, to migrate to urban areas in search of better employment opportunities. Although there are many young people who desire to experience life in urban areas (Seyfrit et al., 2010) there are also many who would prefer to live within a close-knit community atmosphere and lead the healthier lifestyle that can be associated with living in rural areas (Grgić et al., 2010). This exodus to the city by younger generations results in them leaving behind communities populated mostly by the elderly leaving few able-bodied youths to perform essential labour, and women being left behind to raise children alone, and in some cases parents being separated from their children, who are left behind to be raised by elderly relatives. Thus, economic instability is often devastating to the social fabric of rural communities, tearing families apart and destroying traditional cultural values. Those that remain in the rural areas often attempt to eke out a living through unsustainable agricultural practices, which can exhaust the very resources they are relying on for survival. As agricultural land becomes depleted of nutrients, and lacking in adequate water, rendering the land infertile, more forest lands are cleared for agriculture. This environmental destruction is not only bad for the local community but also has a global impact, diminishing biodiversity and contributing to the accumulation of greenhouse gases in the atmosphere.

Agriculture is the main source of income for many rural communities, and issues relating to this will be discussed in more detail in Chapter 4. However, agriculture can be a difficult and unreliable way to make a living. Harvests are seasonal and are frequently adversely impacted by poor weather, disease and other natural phenomena. Pest invasions in crops or outbreaks of disease in livestock can very quickly decimate a family's food supply and annual income. To achieve economic stability, it is often important for families in rural areas to complement agricultural activities with other more resilient sources of income to help lessen this risk. For example, Nathan and Rosli (2014) found a positive correlation between sole reliance on income from agriculture and poverty in rural Malaysia, where the non-agricultural share of household income is becoming increasingly important to a family's survival. In this chapter we describe opportunities for rural communities to generate a more consistent income, without destroying their natural resources, thereby ensuring their children and grandchildren will also be able to provide for themselves.

Figure 2.1 Tea harvesting in Assam, India

Figure 2.1 depicts tea harvesting in Assam, India where the natural environment is conducive to the growing of tea and thus, the local community have responded to this by developing a tea growing industry for making a living. The tea industry in Assam not only provides opportunities for jobs and income for the rural communities but also defines cultural heritage of the particular tea-garden community across the state.

While an agriculture-based economy is prevalent in rural communities in most developing countries, understanding the local trends and potentials of agriculture among other income generating opportunities requires evidence-based enquiry. For instance, on the island of Majuli, India, a survey has been conducted by a team of researchers from the University of Melbourne to determine the needs of the villagers. In that survey respondents were asked about their occupations and livelihood.

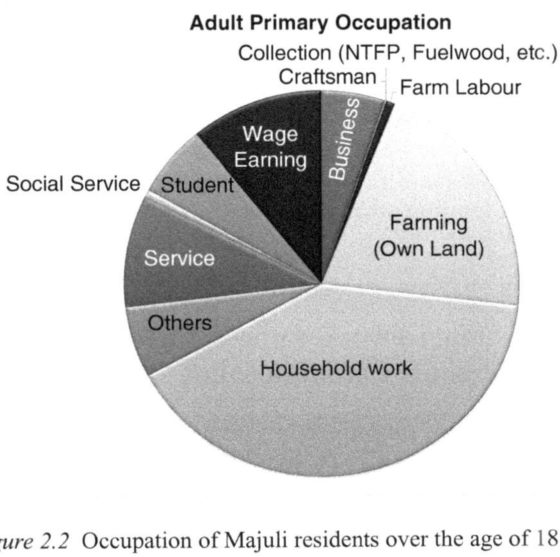

Figure 2.2 Occupation of Majuli residents over the age of 18

Figure 2.2 shows the distribution of income generating activities of the residents of Majuli Island (excluding those under the age of 18, who are mainly occupied as students), which shows household work and farming to be the most commonly held occupations.

The exploitation of forests for financial gain is usually associated with timber production; however, forests contain many other resources that could be harvested in a less destructive way. The second section of this chapter describes some of these opportunities.

Tourism is another form of income generation that has the potential to provide multilevel wealth generation, as locals can offer accommodation, run restaurants, work as tour guides, sell souvenirs and handicrafts, to name but a few of the associated businesses that can be generated by tourism. Tourism can, however, have some very negative social impacts and on local environments (Green, 2005). It is, therefore, important that it be implemented in ways that attempt to avoid these negative impacts. One potentially worthwhile form of tourism is ecotourism. By adhering to the principles of ecotourism, more environmentally and socially sustainable outcomes can be obtained. Unfortunately, ecotourism can also take the form of resort style developments, where accommodation, restaurants and entertainment are all contained within the facility that do not adopt principles of sustainable development. This can not only deprive the local community of the income generation potential but can also destroy their local environment and the very features that attract tourists in the first place and add extra stress on poorly maintained infrastructure and services. In the third section of this chapter, examples of ecotourism are discussed in more detail, including ways it can be implemented to promote more sustainable development of rural communities. Lack of financing availability is often the biggest roadblock to a rural community's exploiting income generating opportunities and in the final section of this chapter innovative models for finance provision are discussed.

Businesses around agriculture

Agricultural activities produce significant amounts of organic waste, which discarded untreated into the environment poses a significant hazard to both human health and the environment. However, if treated through composting the waste can become a valuable source of soil nutrients, reducing the need to purchase synthetically manufactured fertilisers. Case Study 2.1 describes a successful business that developed through composting other people's waste.

Case Study 2.1 Camperdown Compost, Victoria, Australia (Personal Communication, Tony Evans)

The Camperdown Compost Company is a successful business with an annual turnover of $AUD 3million and a staff of 16. There are three components to the business. The first and main component is a large composting site which takes in many typically difficult to treat waste streams including waste from dairy farms, abattoirs, egg hatcheries, ice cream manufacturing, grease traps from restaurants etc. The waste is currently treated in windrows that are 4 to 5 metres high and periodically turned using an excavator to prevent them turning anaerobic. The process is operated continuously and it takes approximately 12 months for the compost to reach maturity.

Over the next three to four years they will be transitioning to incorporate a self-propelled turner. For this to work the compost piles need to be reduced to 2 metres, however, the time to reach maturity will be reduced to around 12 weeks, and the process will convert to a batch system.

Analysis of the mature compost shows it has high quantities of many of the nutrients that are often supplied by synthetic and mineral fertilisers, although the company can adjust the nutrient content to order by adding supplements. Even with bespoke nutrient adjustments the compost is still significantly cheaper than purchasing manufactured fertilisers, while having the advantage of a high carbon content helping to increase the quality of soils.

The price of the compost has decreased significantly over the years from around $AUD 100 per tonne, to its current price of $AUD 10–25 per tonne (the variation depends on the size of the order). The customers tend to be large farms that purchase compost in bulk, rather than nurseries or residential customers. The reason the profitability of the business has grown is due to the rising cost of landfill gate fees, which has allowed the business to charge more for receiving the waste. Landfill fees have been increasing to deter biodegradable wastes from landfills where they can lead to environmental hazards such as GHG emissions, odour nuisance, fires and explosions and local air quality pollution. The company works closely with the Environmental Protection Agency (EPA) to ensure their process minimises emissions.

The second component of the business is assisting large dairy farms to establish their own in-situ composting. This allows them to treat their waste in an environmentally safe way, save them expensive waste collection fees, and allows them to produce their own fertiliser product.

The third component of the business came about as other types of farmers were interested in producing their own compost but were not producing the same level of biodegradable waste as dairy farms. In this case they developed a system where farms would take in locally produced waste from other sources such as residential green waste collections, restaurants etc. Waste would be delivered to the farms where the farmers would process and use the waste for their own purposes.

Similar to the Cristalino Park Reserve, the company was started by a single person, Tony Evans, and took around three years before starting to turn a profit. It continues to be a largely family run enterprise with Tony's wife and eldest daughter both working in the business.

Figure 2.3 Indian market selling coloured powder for the "Holy" festival

Assisting farmers to increase their yields also has the potential to become a profitable business. For example, in the Sahel region, at the convergence of Burkina Faso, Niger and Mali, groups of young men are visiting farmers to teach them a technique known locally as Tassa or Zaï planting (Pretty et al., 2011). The technique involves digging a series of pits by hand that are 25 cm in diameter, 20 cm deep at intervals of 90 cm. Fertiliser is placed in the hole followed by a plant. The shape of the hole prevents run-off ensuring that nutrients and moisture are concentrated at the roots of the plants. Building a stone wall around the field can divert rainwater back towards the Zaï pits, further increasing moisture retention (Critchley & Prima, 2014). The technique not only improves yields in existing crop lands but also has been shown to rehabilitate degraded land, which is encouraging farmers to purchase and cultivate degraded lands. An estimated 3 million hectares of land have been rehabilitated via this technique. Groups of young men are able to travel from village to village working with farmers to cultivate their lands, allowing them to generate an income and preventing them from migrating to urban areas.

Days of cultural significance can also provide a unique income generating opportunity in some regions. For example, in India, during festival times such as Diwali and Holy, people will purchase speciality items used during the celebrations. Figure 2.3 for example illustrates pop-up markets selling the coloured powder used during the "Holy" festival.

Non-timber forest products

Harvesting timber is the main way forests are exploited for financial gain. However, the need to continually increase profits often leads to timber being harvested at a greater rate than it can be replenished, eventually resulting in the loss of the resource. It is either a limited or short-term source of income and thus unsustainable. Harvesting timber can also be culturally insensitive as many forest dwelling communities have an innate desire to preserve the forest for spiritual reasons (Hernández-Aguilar et al., 2017, Esler, 2016, Choy, 2018). Forests contain many other assets that can be exploited in a non-destructive way. In India the leaves of the Palash tree are used to weave disposable plates, which can then be used instead of plastic plates to serve food on at markets (Chavan et al., 2016). In Southern Mixteca, Mexico pine resin is extracted and used to make turpentine (Hernández-Aguilar et al., 2017).

Reforestation commitments that form part of the recent Paris Agreement on climate change opens up potential income generating opportunities. For example, in Brazil, the lack of native seed species availability is a potential hindrance to achieving the Atlantic Forest biome reforestation commitment (Silva et al., 2017). Nurseries are the main source of seeds and current stocks are inadequate, leading to the temptation to plant cheap, readily available alien species. However, the ease with which these species grow generally indicates they will be invasive and could cause loss of biodiversity in nearby virgin forest areas. The collection of native seeds from existing forest areas is an income generating opportunity that could avoid this situation. Silva et al. (2017) noted that freelance seed collectors are making up to $US 1400 per month, four times the current minimum wage in Brazil. Through training forest dwelling communities on how and which types of seeds to collect significant income could be generated while at the same time complying with climate change commitments and preserving biodiversity.

Niche commodities are another income generating opportunity for rural communities. Niche commodities are specialised products sold to a limited number of customers at a premium (Waroux & Lambin, 2013). An example is the silk and muga industry in Assam, India depicted in Figure 2.4. This is a highly specialised product where both upstream and downstream processes are managed in a very traditional way (see Case Study 2.2). The industry is supported by the local market where the woven fabric is utilised in numerous cultural functions such as weddings. There are five broad niche commodity markets: environmental concern (organic, sustainable harvesting); responsibility to others (fair trade label); women's empowerment; perceived health or cosmetic benefits; gourmet foods. Niche markets have the potential to alleviate poverty in a sustainable way, as was the case with acai and goji berries. However, this is not always the case. Waroux and Lambin (2013) investigated whether argan oil was leading to poverty alleviation in Morocco. They found that while women from poor rural communities were performing the extraction of the oil, they were receiving only a minimal wage for doing so. When asked what the market value of the oil was, the women did not know. The majority of the profits from oil sales were retained by export companies, despite the product being marketed as contributing to women's empowerment and alleviating poverty.

Case Study 2.2 RGVN Assam, India (Personal Communication, Amiya Kumar Sharma & Rachna Yadav)

Rashtriya Gramin Vikas Nidhi (RGVN) is a non-governmental organisation that conducts infrastructure projects for rural communities. Other organisations often set a limit on the level of assistance that they will provide, for example many state they will spend three years working with a rural community before moving on to the next project. RGVN's philosophy is to continue working with the community until they prove they are self-sufficient.

An example of the good work they have been doing is their work with a forest community in Assam, India to establish a textile factory. The factory produces artworks using traditional craft skills of the region. Thus, they are bringing economic prosperity to the village, while keeping their cultural traditions. They worked with a group of people from the village for one year during which time the group mastered the craft skills and were producing a steady output of artworks. While it may seem that this has brought self-sufficiency to the community, currently the artworks are only sold locally with limited scope for expansion. Therefore, RGVN intends to continue working with the group to help them develop marketing skills to allow them to expand their customer base. The aim is to continue to grow the textile business allowing it to employ more people, bringing improvements to the whole community. Their hope is to leave the village generating an income sufficient to meet their basic needs, without damaging the forest, and while preserving their cultural traditions.

Figure 2.4 Weaving silk in Assam, India

Thus, while forests offer many opportunities for small and medium enterprises (SMEs), there are many obstacles to their success. Access to financing and markets, policy and regulations, and achieving economies of scale were highlighted by Hajjar et al. (2017). Another issue is that forest dwelling communities often do not own the land in a legal sense. By exposing the potential profitability of forest products, they may be vulnerable to large companies purchasing the land from the government and bringing their own workers in to perform the harvesting (Chavan et al., 2016). If governments are committed to achieving sustainable development goals, particularly alleviating poverty, they must offer forest dwelling communities some ownership of the land they occupy to prevent their livelihoods from being sold out from under them.

Ecotourism

Ecotourism is defined as responsible travel to natural areas that conserves the environment and improves the well-being of local people (Bluwstein, 2017, Mayett-Moreno et al., 2017). Further to this, many "ecotourists" express a desire to learn about and experience local traditions and customs as part of their overall travel experience. By incorporating cultural experiences into an ecotourism development project, the three pillars of sustainability can be adhered to: environmental conservation is achieved in a way that promotes economic development through job creation and increased income generating opportunities; social development is created through improved services and infrastructure, while preserving cultural traditions. However, the reality often falls short of the ideal. In this section we will look at each of the three pillars of sustainability separately and examine ways that ecotourism ventures can fail to achieve the ideal and discuss how they could be better implemented.

Economics

Poverty alleviation and nature conservation have often been considered competing, rather than complementary goals for rural communities (Yergeau et al., 2017, Dedeke, 2017, Jolin, 2018). However, ecotourism can provide a positive correlation between these two goals (Yergeau et al., 2017). Theoretically, ecotourism should increase economic diversification, foreign exchange earnings retention and employment opportunities, promote the development of infrastructure, and create industrial linkages and opportunities for low-skilled labourers and women (Chirenje, 2017). The development itself will create jobs, while the increased number of visitors to the area will create an opportunity for SMEs to flourish. In reality, profits are often retained by tourism companies from developed nations; chain hotels and restaurants open up in the area drawing business away from local SMEs (Supradist, 2004); the lack of education and skills among locals drives developers to employ outsiders. Chirenje (2017) reported that only 11 per cent of the staff at an ecotourism development in Zimbabwe were locals, only half of these were full time and they were in the lowest level jobs, receiving less than minimum wage. Ecotourism developers in the Waterberg Biosphere Reserve in South Africa expressed reluctance to waste time and money training staff to fulfil newly created employment, preferring to bring in outside staff that already had relevant training and experience (Lyon et al., 2017).

Despite their lack of education locals have knowledge of the surrounding area that could provide a valuable resource for establishing an ecotourism industry. When exploring a vast area of native forest, locals have the knowledge of the best places for sightseeing, fishing, hiking etc. Exploiting this knowledge to enhance the tourist experience could be very valuable. Idris et al. (2016) provided locals with GPS enabled mobile phone, so they could document the best locations for tourists, using a simple app. Although difficulties were faced when the app required extensive user input, as many of the locals were only semi-literate and/or had little or no experience with mobile phone technology, using an app with icons or some other type of interface proved very successful.

Whether or not an ecotourism venture boosts the local economy often depends on the way it is developed and implemented. A top-down approach, where the government initiates tourism activities without consultation of the local people will generally result in a failure to positively impact the local economy (Orenstein & Shach-Pinsley, 2017). Often, land will be sold to an international corporation, and all profits from the tourist trade will go to them. Not only do these developments fail to provide income to locals but also in a few cases locals have been driven off their land to make way for the new development (Bluwstein, 2017). A bottom-up approach, where tourist activities are instigated by the local community is much more in keeping with the Smart Village concept as all profits will be retained by the village. However, few rural communities are likely to have the knowledge and resources to provide all the services tourists expect (Orenstein & Shach-Pinsley, 2017). They also lack other essential skills such as the ability to come up with an effective marketing campaign to attract tourists to the region. Vitória Da Riva Carvalho, for example, attempted to engage travel agencies to help promote her ecotourism development in Brazil, but was rejected as the companies did not trust her lack of experience (Dedeke, 2017) and were concerned about the quality of infrastructure surrounding the development (see Case Study 2.3 for more details on this). The lack of government involvement could also hinder a tourism development as they are required to provide essential infrastructure. For example, poorly maintained roads could physically prevent tourists from accessing the region.

Case Study 2.3 Vitória Da Riva Carvalho and the Cristalino Private Forest Reserve (Dedeke, 2017)

Vitória Da Riva Carvalho founded an ecotourism industry in Mato Grosso, Brazil that led to the designation of 184,900 hectares of Amazonian forest as protected land. She was following in the footsteps of her father who purchased 80,000 hectares of forest in the mid-1970s with a dream of establishing a sustainable agricultural community. He divided up his land among 150,000 people creating the town of Alta Floresta. Unfortunately, a drop in the price of commodities, coupled with a gold rush at the start of the 1980s saw his dream shattered. The agricultural venture became economically unfeasible, driving his community to more intensive agriculture and mining which resulted in the destruction of much of his forest.

In 1990 Vitória purchased 700 acres of forest in the same state with a vision of establishing an ecotourism business. She had built an ecolodge by 1992, however, faced various obstacles to success. Her attempts to engage the local community failed, as they saw her idea as a threat to their agricultural livelihood and did not believe it would generate sufficient income. She contacted tourism companies asking them to promote her lodge but they were not interested as the area was not well known for tourism and had poor transport infrastructure. Undeterred, Vitória studied business and tourism and set up her own tour company, promoting the lodge herself. At first business was slow until a renowned ornithologist, Ted Parker, visited the region and discovered several rare and unknown species of birds. Publication of his findings prompted other bird enthusiasts to visit the area, which sparked the interest of international tour operators. More scientists were attracted to visit the region for research purposes including representatives from the Royal Botanic Gardens, Kew, UK. They found a record nine never before seen habitats. Previous studies from much larger forest areas had only found one or two. Their research confirmed the scientific significance of the area and by 1998 the state government designated 66,900 hectares as protected. Vitória founded a non-profit institution, Cristalino Ecological Foundation in the area to continue to encourage visitors to the region and further the important scientific discoveries. By 2001, the size of the designated area was expanded to 184,900 hectares and partners of the lodge attempted to engage the local community to be a part of the industry to ensure their support. However, this was only partially successful and local government attempted to overturn the designation of the area, freeing up almost half of it for development, multiple times. Fortunately, local conservationists with backing from the international scientific community finally saw the protected status of the area confirmed in 2010.

Around 50 of the 515 bird species observed are endemic to the Cristalino State Park, highlighting the significant preservation of biodiversity resulting from the protection of the area. Vitória's actions demonstrate that ecotourism can be an effective way to make environmental conservation profitable.

The two approaches described so far consider four groups of tourism development: stakeholders, tourists, local residents, developers and government and how their conflicting interests prevent the economic ideals of an ecotourism venture from being realised. An NGO can provide an important intermediary between the other stakeholders (Lyon et al., 2017). They would allow the expertise of a corporation to manage the development, maximising its chances of success, while at the same time ensuring local people's interests are protected. This provides a middle ground between a top-down and bottom-up approach. For example, the government is allowed to sell land to a corporation but are required to use at least part of the money to develop or upgrade infrastructure to improve life for locals as well as making the area safe and appealing to tourists. The corporation is allowed to develop the land but must be required to employ a set percentage of their staff from the local community, and to provide associated training to ensure they are not only given the lowest paid jobs. The development must also encourage tourists to explore the local community and support their businesses.

Social

By creating employment opportunities in rural areas, ecotourism could reverse the tide of urban migration, reuniting young people with their families. However, it may be difficult to convince people of its potential. In Mexico, Mayett-Moreno et al. (2017) interviewed young men who returned to their home village for the implementation of an ecotourism venture. Many had worked in the hospitality industry during their time away, acquiring relevant skills. However, the participants believed the job opportunities were limited in that they would only pay minimum wage and be temporary or seasonal. They did not believe they would earn a sufficient income to meet their basic needs. This inability to see the potential for growth from the ecotourist venture further highlights the need to include the local community as a key stakeholder at all stages of development.

Preserving cultural traditions is also an important component of the social pillar of sustainable development. While ecotourism could be viewed as an antithesis to this, by bringing external influences into an isolated community, it could provide a unique opportunity to not only preserve but also enhance local culture. Various studies that set out to identify the attraction of ecotourist destinations have found that experiencing a new culture ranks very highly (Lee & Hsieh, 2016, Dillette et al., 2017). Abdurahman et al. (2016) noted that tourists expected to participate in traditional cultural activities, such as dancing and handicrafts. Traditional cooking courses are also very popular.

Unfortunately, corporations sometimes may have a preconceived notion of what cultural traditions will look like and may try to force this image onto the local community, rather than approaching them to learn about their traditions. Lyon et al. (2017) noted that in the Waterberg Biosphere Reserve in South Africa, locals feel they are being brought in to dance for the white man, in a throwback to colonial times where locals were a spectacle to be gawped at. Similarly, Bluwstein (2017) noted that real locals may be moved on from where they are living, to hide them from the tourists so they don't spoil the fetishised version of the "traditional lifestyle" the tourism developers want to create. It is important to respect that local cultural traditions belong to the local people and it is their choice when and how these are to be shared with outsiders. Any profits to be made from sharing these traditions should be retained by the local community.

Improved infrastructure and associated services are often cited as a beneficial by-product for local communities due to tourism development, particularly in the case where they are not directly profiting. However, this may not necessarily be the case. In parts of Spain, for example, overdevelopment of prominent tourist areas led to the destruction of the natural beauty that attracted tourists to the area in the first place (Bickford et al., 2017). Anctil and Blanc (2016) simulated the impact of a tourism development in a coastal area with no formal waste or wastewater treatment infrastructure. In the baseline situation the small size of the local fishing community meant that waste and wastewater were absorbed into the environment without creating excess damage. After the tourism development, the increase in effluent was shown to eventually lead to loss of fish stocks, destroying the livelihood of the local community. Eventually, the water sources would become so polluted that the site would no longer be an attractive tourist destination. Thus, while the site developers receive a quick return on their investment, the development would not be profitable long-term. Thus, developers must be required to invest in infrastructure to not only protect the livelihoods of locals but also to ensure longevity of their investment.

Volunteer tourism is a subset of ecotourism, in which tourists aid local development. One of the main motivations of the volunteer tourists is to experience a new culture (Dillette et al., 2017). These types of tourist organisations can be a more effective way of improving livelihoods. To be effective, the volunteer groups should not just come in and build infrastructure for the community, but rather work together with community members, giving them the skills to build and maintain infrastructure, after the tourists have gone home. Volunteers could also teach the local community about business and marketing, helping them to increase the profitability of their tourism venture. The important thing is to ensure that the volunteers are bringing expertise in areas relevant to the betterment of the local community, so that the outcomes for locals who participate in the programme are long-term improvements in their livelihood.

Environment

Wildlife often forms the main attraction at an ecotourism site (Lyngdoh et al., 2017, Bateman & Fleming, 2017). Thus, the preservation of fauna species and the ecosystems in which they thrive is paramount to the success of the venture. The local community may naturally desire to preserve these ecosystems due to their cultural beliefs. Tibetan Buddhists, for example, believe there are spirits living in all components of the earth and by destroying these we will incur their wrath (Esler, 2016). Interviews with a forest dwelling community in Malaysia found a strong sense of responsibility to preserve the forest for future generations, even though they had never heard of the concept of sustainable development (Choy, 2018). On the other hand, locals may feel that exploitation of their natural environment, for example harvesting wood to burn for energy generation or cultivating land to grow food, as essential to their survival. Therefore, it is important to ensure the local community has a sense of responsibility towards ecotourism by ensuring they are profiting from the venture.

Concerns have also been raised about the impacts of ecotourism on the animals that form their main attraction. Bateman and Fleming (2017) compiled data from 300 studies into animal responses to increased human presence. Their interpretation of the results showed that animals became used to the presence of humans over time and were less likely to flee. From an ecotourism perspective, this is a doubly positive result as it means that many species of animals may not be stressed by the increased presence of humans, and that tourists have increased chances of seeing them. The only potential downside is the lack of wariness could make the animals more vulnerable to poachers, or other illegal activity that would endanger them. The local community will play a key role in protecting wildlife from these sorts of threats, further emphasising the importance of their involvement in the venture.

While high biodiversity is often seen as a priority for ecotourism, to be considered truly sustainable the initiative must incorporate the full spectrum of environmental management measures including minimisation of greenhouse gas (GHG) emissions. Yet, this is often overlooked, or incorporated in a superficial way. For example, many countries push to make a small island tourist destination have 100 per cent renewable energy for electricity production (Grydehøj & Kelman, 2017). While this claim looks good on an advertising poster, the GHG savings can often be disproportionate to the cost of implementing the programme. Spending the same amount of money to develop renewable energy infrastructure on the mainland would impact a larger proportion of the population, leading to much more significant GHG savings.

Similarly, many regions emphasise incorporating "visible" environmentally friendly measures, such as solar panels (Grydehøj & Kelman, 2017). A Life-Cycle Assessment (LCA) of these places often finds the benefits are at best marginal. For example, while the electricity generation may come from 100 per cent renewable sources, transport to and around the location still uses traditional fossil fuel burning vehicles leading to a net increase in GHG emissions. Transport and energy generation will be the main sources of GHG emissions for most tourism ventures and efforts must be made to reduce GHG emissions from both of them. Transport options around the site will be easier to control. Bicycles are a good carbon free option where paths are available. Electric vehicles may be the most appropriate choice, if adequate electricity provision is available and the source of energy is renewable and non-polluting. Animal transport can also be a good option and tourists may enjoy the authenticity of this experience.

As we will see in Chapter 8, water resources are stretched in many areas. The introduction of tourism could add further strain. Lee and Wang (2017) surveyed the local community in an up-and-coming ecotourist afforestation area of Taiwan and found water quality one of the main concerns relating to the project.

Efforts should be made to make the tourism site as self-sufficient as possible for example, through the use of rainwater collection tanks and water purification equipment, which will provide a safe source of drinking water for tourists, while preventing the need to purchase wasteful plastic water bottles. Tourism accommodation must also treat wastewater prior to releasing it into the environment to prevent negatively impacting local livelihoods and impairing the tourist appeal of the location (Anctil & Blanc, 2016).

Tourism in any form can, however, have both negative social and environmental impacts on the host communities, as illustrated by previous research (Green, 2005).

Financing opportunities

There are two barriers to impoverished members of rural communities exploiting business opportunities, such as those described so far in this chapter. The first is lack of awareness of the potential for generating income from these activities. The second is lack of access to the finance required to establish and operate a business. Some of the main reasons people in rural areas have difficulty accessing finance services include (Allison et al., 2015):

- lack of physical presence of a bank in their area;
- difficulty in reading and filling in required paperwork for a loan application;
- unwillingness of banks to provide loans to people from rural communities. Many banks consider the risks too high, and prefer to invest with more established businesses, people with assets that could be recovered in lieu of loan payments, or businesses in urban areas where the perceived customer base is more visible (Personal Communication, RGVN).

Other routes to obtaining start-up capital are also inaccessible to poor rural communities such as personal savings, which can be difficult to accrue where there is no bank to deposit cash and help with the saving process. Similarly, many households' monthly income barely covers their monthly expenses, let alone leaving extra money to put aside.

Informal borrowing from relatives or friends is only an option if such a person is available who has money to spare. Informal loans from other community members are also possible, however, the lack of regulation to protect people who take out these loans leaves them vulnerable if they are unable to repay the loan. There is also no control over the interest rates of these loans and they can be extortionate.

Microfinancing was conceptualised to overcome these difficulties. The "birth" of modern microfinance is considered the establishment of the Grameen Bank, in Bangladesh in 1975 (Allison et al., 2015). The theory goes that providing a small loan will allow a small business to get started, without accruing excessive debt. The loans target the poorest members of the community and do not require any collateral. The interest rates are very low and a generous time allowance for repayment is made.

Early studies indicated that microlending was an effective tool for poverty reduction, that the effectiveness was more pronounced for female than male borrowers, that the extremely poor benefited the most, and that microcredit enables consumption smoothing. Due to the reported success of the initiative, in 2006 Muhammad Yunus and the Grameen Bank were awarded the Nobel Peace Prize for their contribution to poverty alleviation (Odell, 2010). However, more recently there have been negative reports of Micro Financial Institutions (MFIs) profiting from the very poor, leaving them with debts that they cannot repay (Banerjee et al., 2015a). This led to a series of more robust studies into the impact of microfinancing on poverty alleviation and sustainable development goals. Banerjee et al. (2015a) documented the impact of an MFI that was set up in Hyderabad, India from its inception. Only around 18 per cent of people offered the loan decided to take it out and these mainly consisted of people who had expressed a desire to take out a loan prior to the product becoming available. The loans that were taken up were mainly to help existing businesses to expand and very few new businesses were established. No improvement in sustainable development goals, such as women's empowerment or educational attainment, resulted from the availability of the loans. Odell (2010) reported on the conflicting results between early and more recent studies into the impact of microfinancing on poverty alleviation. One of the reasons for the discrepancy was a lack of robustness in the methodology of the earlier studies. Overall, they found that while the loans didn't appear to be aiding sustainable development goals, mean household incomes had increased. There were also some confounding factors, for example a small family business may have employed staff to fill a role that was previously performed by children, so that the children could attend school. Overall, the household's income would appear to be the same or less than before, even though their situation had improved from a development perspective.

Several articles suggest that simply making microfinancing available may not be enough to support a family out of poverty, people also need skills and knowledge to transform their situation. Pretty et al. (2011), for example, noted many situations where a farming community worked hard to increase their yields, only to fail to be able to sell the surplus stocks leaving them no better off than they were before. Many other studies have noted that small loans were often used to purchase food or other household items, rather than to invest in a business (Attanasio et al., 2015). Providing a microfinancing programme along with training in business development was projected to be a more effective approach to poverty alleviation. Banerjee et al. (2015b) implemented a six-component, two-year targeted programme in six countries that included:

1. provision of an asset that could be used to establish a business (such as chickens),
2. a monthly allowance (to prevent the family simply selling the asset to pay for consumables and ending up once again asset-less),
3. a training programme in how to effectively develop the asset into a business,
4. regular household visits to check up on progress,
5. access to a savings account (which in some instances had a compulsory monthly deposit but not all),
6. training in another life skill, mostly related to health.

The programme targeted the poorest households in the poorest villages. Participating households were surveyed prior to programme implementation, at the end of the programme, and one year after completion to assess whether the improvements had been sustained. Most households achieved income improvements by the end of the programme, which was maintained in the follow-up survey a year later, but they did not achieve further income growth. Thus, while the household's situation improved, they remained comparatively poor. The survey also looked at a range of developmental improvements. All of these were maintained at the final survey, except for women's empowerment. Although successful in many respects, the costs of programme implementation are too high to be feasible on a broader scale. However, it may be that implementing only some of the steps outlined above could be just as effective while much more economically efficient. The regular household visit was noted as a particularly expensive component, the omission of which would make the programme more realistic to implement. In South Sudan the "Star Support Group", a charitable organisation, decided to try microfinancing (Personal Communication, Gerard Ellul). They loaned 6000 South Sudanese pounds (SSP) (just under $US 50) to 61 people with small business ideas, accompanied by some training. Within 6 months, 38 had repaid in full, while 22 had partially repaid. Only one loan was not repaid, as the loanee passed away shortly after taking out the loan. The success of the scheme allowed the Star Support Group to offer a further loan of 10,000 SSP (approximately $US 77, $AU 100) to those who had repaid, and to expand the initial loan of 6,000 SSP to more people.

Spreading awareness of a loan scheme can also be a challenge in remote rural communities where communication is dominated by word of mouth rather than other media. Banerjee et al. (2013) explored which individuals in the community to target to maximise awareness of loan availability and loan uptake. The experiment surveyed communities to identify individuals that were most likely to have a strong influence (teachers, village leaders etc.) and targeted these individuals with an information session about the loan scheme. They then tracked how information spread from these to other members of the community. The assumption was all information dissemination would be within a village, and not between adjacent villages. They found that people who decided to participate in the scheme were more likely to pass on information about it to others. There was no difference in uptake between those who were informed about the scheme by participators than non-participators. Encouraging individuals to spread the word about the loans, regardless of whether or not they decide to take on loans could be an important way of reaching the broader community.

Early on, microfinancing was provided by charitable organisations; however, as demand for loans increased the NGO model became untenable for the scale of management required (Battilana & Dorado, 2010, Hermes et al., 2011). This coupled with the fact that a substantial history of successful microfinancing schemes now exists has provided a model for establishing microfinancing as a commercial venture. For the major banks (HSBC, Deutsche Bank etc.) there is a twofold benefit – it gives the impression of participating in corporate social responsibility, while still generating a profit (Hermes et al., 2011). As microfinancing

NGOs transition into commercial enterprises concerns have been raised about their ability to retain the social responsibility of targeting the poorest members of the poorest communities. Finance experts are inclined to alter their businesses to maximise profits and may be tempted to reassess the criteria for loanees to already established businesses looking to expand or to wealthier clients who have assets to put up as collateral. Financiers often claim that by allowing existing businesses to expand the rest of society will benefit in a trickle-down effect – bigger businesses will employ more staff thereby creating jobs and reducing unemployment. However, as we discussed in the ecotourism section, business operators are inclined to hire workers with relevant qualifications and experience and would rather bring in skilled workers from outside their community than employ locals who would require substantial training to perform the job requirements. Conversely, staffing a microfinancing business with social workers could lead to a failure to maintain profitability.

Battilana and Dorado (2010) described two microfinancing NGOs that transitioned into commercial enterprises – Los Andes and BancoSol. BancoSol employed a mixture of financial experts and social workers, guided by the philosophy that the two groups would rub off on each other, helping the business to stay focused on the twin goals of improving social well-being, while generating a profit. Although the company was initially very successful, over time the two groups of workers, rather than influencing each other, found it more and more difficult to reconcile their differences. The finance experts began to push out the social workers and the company became a more traditional loan organisation requiring collateral. Los Andes, on the other hand, decided to only employ recent graduates, based on the philosophy that they would provide a *tabula rasa* allowing them to develop their work experience with both goals simultaneously. As a result, the company took longer to become established as many of the staff was learning on the job. However, over time the model proved more viable and the company has managed to retain its goals of social development while becoming a profitable business venture.

The motivation behind the entrepreneurs who start up an MFI could strongly influence its ability to achieve the dual goals. Wry and York (2017) described three types of social entrepreneurs: single-minded who are focused on a single goal that is either making a profit or "doing good". These types build a business based on an existing model with a proven track record. Mixed entrepreneurs have multiple goals to a different degree. For example, they may come from a finance background giving them the skills to identify and establish a profitable business. However, they may have a personal moral desire to use their entrepreneurial skills to "do good". Similar to single-minded entrepreneurs they tend to follow an existing business model and will either sacrifice profits in order to fulfil their desire to improve social development or compromise their social development to retain profitability. If they receive excessive criticism they are likely to abandon their enterprise and try to satisfy their conflicting goals in other ways. For example, they may establish a traditional profitable business, and do volunteer work in their spare time. Balanced entrepreneurs, similar to mixed ones, also have desire to both establish a profitable business and do good but are less prone to one or the other. Therefore, rather than compromising they will redevelop a business model to align their two goals. Allison et al. (2015) explored the motivation of investors on an internet-based crowdsourcing fund by correlating business descriptions, the time taken for the loan amount to be reached, and various factors related to how the business was described. The site does not require any collateral or offer any compensation for repayment failures, although the size of the loans is relatively small. These factors included extrinsic descriptions, such as potential profit margins as in a traditional funding bid, and intrinsic descriptions, which emphasise personal information about the business owner and their situation. Investors were found to be more readily motivated by intrinsic factors, indicating potential investors accessed the site with a desire to "do good" and were more concerned about the potential improvement in well-being their investment would have, rather than making a profit. Thus, the desire to do good can override the desire to maximise profits, however, reworking a business model to align the goals more closely would optimise the value of an MFI.

Once the loans have been taken out, the MFI has no control over what they are used for and various instances have noted that, particularly where loans are targeted to women, a male relative will get her to take out the loan and then take the money from her. In other instances, the loans are largely used to pay for groceries and household items, rather than invested in a business. One suggestion for overcoming this obstacle is to make loans to groups rather than individuals. On the one hand, group members may feel a greater sense of responsibility to pay off their share of the loan as otherwise the whole group will be negatively impacted. On the other hand, they may feel that the group will "cushion" them and cover their share of the loan, to avoid their own personal loss. Attanasio et al. (2015) performed a comparative study between group and individual loans offered to women in Mongolia and found that group loans were more effective at establishing businesses while individual loans were more likely to be used to cover household expenses or loaned on to another relative. In terms of repayment, however, both groups performed similarly. Loans were repaid faster by repeat borrowers or people with greater assets. Hargreaves et al. (2011) suggest that providing loans with conditions, such as requirement to attend a health clinic or send children to school, could be an effective way to ensure the loans support attainment of development goals, although this could be difficult to monitor and enforce.

Recommendations for best practice

The ability to generate an income sufficient to cover basic needs is the key to poverty alleviation. It is important that any new business ventures are conducted in a way that preserves the natural resources the community is relying on for survival, to ensure that rural communities will be self-sustaining in the long-term, not just temporarily.

Extraction of forest products that do not lead to destruction and loss of biodiversity are often a good option. Ecotourism ventures are another good opportunity to provide income while preserving resources. To be able to take advantage of both of these approaches to income generation, communities will need to be provided with training programmes to develop new skills, not just in the activities described, but general business skills, such as marketing, that will be essential to making any business venture a success.

Financing is essential to help businesses become established and grow. The Grameen Bank microfinancing model can provide the basis for this by offering small loans that require no collateral, with low interest rates and long payback periods. Until now, the success of these loans has hinged on the fact they were offered by NGOs with no desire to generate profits but as demand for loans increases and NGO microfinancing programmes are forced to transition to commercial enterprises, the risk that the desire to generate profit will outweigh the desire to alleviate poverty is concerning. Staffing the enterprise with a mix of experienced financiers and social workers was not found to be the most successful approach, compared to staffing the company with inexperienced recent graduates. Finding entrepreneurs with balanced goals seems to be the key to ensuring the programme management is committed to both goals of alleviating poverty and maximising profits. These types of entrepreneurs are most likely to evolve traditional microfinancing models into their most effective form, rather than relying on emulating current models.

Making loans available is only part of the solution, ensuring they are used to generate businesses, and not just to cover expenses is also a challenge. Loaning money to groups rather than individuals appears to be more successful, similarly requiring loanees to participate in sustainable development improvements, such as attending health clinics or sending children to school could also make microfinancing schemes more effective at driving communities towards attaining sustainable development goals.

References

Abdurahman, A. Z. A., Ali, J. K., Khedif, L. Y. B., Bohari, Z., Ahmad, J. A. & Kibat, S. A. 2016. Ecotourism product attributes and tourist attractions: UiTM Undergraduate Studies. *Procedia – Social and Behavioral Sciences*, 224, 360–367.

Allison, T. H., Davis, B. C., Short, J. C. & Webb, J. W. 2015. Crowdfunding in a prosocial microlending environment: Examining the role of intrinsic versus extrinsic cues. *Entrepreneurship Theory and Practice*, 53–73.

Anctil, A. & Blanc, D. L. 2016. An educational simulation tool for integrated coastal tourism development in developing countries. *Journal of Sustainable Tourism*, 24, 783–798.

Attanasio, O., Ausburg, B., Haas, R. D., Fitzsimons, E. & Harmgart, H. 2015. The impacts of microfinance: Evidence from joint-liability lending in Mongolia. *American Economics Journal: Applied Economics*, 7, 90–122.

Banerjee, A., Chandrasekhar, A. G., Duflo, E. & Jackson, M. O. 2013. The diffusion of microfinance. *Science*, 341.

Banerjee, A., Duflo, E., Glennerster, R. & Kinnan, C. 2015a. The miracle of microfinance? Evidence from a randomized evaluation. *American Economics Journal: Applied Economics*, 7, 22–53.

Banerjee, A., Duflo, E., Goldberg, N., Karlan, D., Osei, R., Parienté, W., Shapiro, J., Thuysbeart, B. & Udry, C. 2015b. A multifaceted program causes lasting progress for the very poor: Evidence from six countries. *Science*, 348, 772–790.

Bateman, P. W. & Fleming, P. A. 2017. Are negative effects of tourist activities on wildlife over-reported? A review of assessment methods and empirical results. *Biological Conservation*, 211, 10–19.

Battilana, J. & Dorado, S. 2010. Building sustainable hybrid organizations: The case of commercial microfinance organizations. *The Academy of Management Journal*, 53, 1419–1440.

Bickford, N., Smith, L., Bickford, S., Bice, M. R. & Ranglack, D. H. 2017. Evaluating the role of CSR and SLO in ecotourism: Collaboration for economic and environmental sustainability of Arctic resources. *Resources*, 6.

Bluwstein, J. 2017. Creating ecotourism territories: Environmentalities in Tanzania's community-based conservation. *Geoforum*, 83, 101–113.

Chavan, S. B., Uthappa, A. R., Sridhar, K. B., Keerthika, A., Hand, A. K., Newaj, R., Kumar, N., Kumar, D. & Chaturvedi, O. P. 2016. Trees for life: Creating sustainable livelihood in Bundelkhand region of central India. *Current Science*, 111, 994–1002.

Chirenje, L. I. 2017. Contribution of ecotourism to poverty alleviation in Nyanga, Zimbabwe. *Chinese Journal of Population Resources and Environment*, 15, 87–92.

Choy, Y. K. 2018. Cost-benefit analysis, values, wellbeing and ethics: An indigenous worldview analysis. *Ecological Economics*, 145, 1–9.

Critchley, W. & Prima, S. D. (Directors) 2014. *Zaï planting pits*. Amsterdam, Netherlands.

Dedeke, A. 2017. Creating sustainable tourism ventures in protected areas: An actor-network theory analysis. *Tourism Management*, 61, 161–172.

Dillette, A. K., Douglas, A. C., Martin, D. S. & O'Neill, M. 2017. Resident perceptions on cross-cultural understanding as an outcome of volunteer tourism programs: The Bahamian Family Island perspective. *Journal of Sustainable Tourism*, 25, 1222–1239.

Esler, J. 2016. "Green Tibetans" in China: Tibetan geopiety and environmental protection in a multilayered Tibetan landscape. *Asian Ethnicity*, 18, 330–350.

Green, R. 2005. Community perceptions of environmental and social change and tourism development on the island of Koh Samui, Thailand. *Journal of Environmental Psychology*, 25, 37–56.

Grgić, I., Žimbrek, T., Tratnik, M., Markovina, J. & Juračak, J. 2010. Quality of life in rural areas of Croatia: To stay or to leave? *African Journal of Agricultural Research*, 5, 653–660.

Grydehøj, A. & Kelman, I. 2017. The eco-island trap: Climate change mitigation and conspicuous sustainability. *Area*, 49, 106–113.

Hajjar, R., Badini, O. S. & Kozak, R. A. 2017. Promoting small and medium forest enterprises in national REDD+ strategies: A global analysis of enabling environments. *Climate Policy*, 17, 731–763.

Hargreaves, J. R., Boccia, D., Evans, C. A., Adata, M., Petticrew, M. & Porter, J. D. H. 2011. The social determinants of tuberculosis: From evidence to action. *American Journal of Public Health*, 101, 654–662.

Hermes, N., Lensink, R. & Meesters, A. 2011. Outreach and efficiency of microfinance institutions. *World Development*, 39, 938–948.

Hernández-Aguilar, J. A., Cortina-Villar, H. S., García-Barrios, L. E. & Castillo-Santiago, M. Á. 2017. Factors limiting formation of community forestry enterprises in the southern Mixteca region of Oaxaca, Mexico. *Environmental Management*, 59, 490–504.

Idris, N. H., Osman, M. J., Kanniah, K. D., Idris, N. H. & Ishak, M. H. I. 2016. Engaging indigenous people as geo-crowdsourcing sensors for ecotourism mapping via mobile data collection: A case study of the Royal Belum State Park. *Cartography and Geographic Information Science*, 44, 113–127.

Jolin, L. 2018. Adventures in life sciences. *Imperial*, 43, 46–47. Imperial College, London: Imperial College, London.

Lee, C.-H. & Wang, C.-H. 2017. Estimating residents' preferences of the land use program surrounding Forest Park, Taiwan. *Sustainability*, 9.

Lee, T. H. & Hsieh, H.-P. 2016. Indicators of sustainable tourism: A case study from a Taiwan's wetland. *Ecological Indicators*, 67, 779–787.

Lyngdoh, S., Mathur, V. B. & Sinhaa, B. C. 2017. Tigers, tourists and wildlife: Visitor demographics and experience in three Indian Tiger Reserves. *Biodiversity and Conservation*, 26, 2187–2204.

Lyon, A., Hunter-Jones, P. & Warnaby, G. 2017. Are we any closer to sustainable development? Listening to active stakeholder discourses of tourism development in the Waterberg Biosphere Reserve, South Africa. *Tourism Management*, 61, 234–247.

Mayett-Moreno, Y., Villarraga-Florėz, L. F. & Rodríguez-Piñeros, S. 2017. Young farmers' perceptions about forest management for ecotourism as an alternative for development in Puebla, Mexico. *Sustainability*, 9.

Nathan, S. B. S. & Rosli, M. M. 2014. Distributional effects of non-farm incomes in a Malaysian rice bowl. *International Journal of Social Economics*, 43, 205–220.

Odell, K. 2010. Measuring the impact of microfinance: Taking another look. *Grameen Foundation Publication Series*. Washington DC: Grameen Foundation.

Orenstein, D. E. & Shach-Pinsley, D. 2017. A comparative framework for assessing sustainability initiatives at the regional scale. *World Development*, 98, 245–256.

Pretty, J., Toulmin, C. & Williams, S. 2011. Sustainable intensification in African agriculture. *International Journal of Agricultural Sustainability*, 9, 5–24.

Seyfrit, C. L., Bjarnason, T. & Olafsson, K. 2010. Migration intentions of rural youth in Iceland: Can a large-scale development project stem the tide of outmigration? *Society and Natural Resources*, 23, 1201–1215.

Silva, A. P. M. D., Schweizer, D., Marques, H. R., Teixeira, A. M. C., Santos, T. V. M. N. D., Sambuichi, R. H. R., Badari, C. G., Gaudare, U. & Brancalion, P. H. S. 2017. Can current native tree seedling production and infrastructure meet an increasing forest restoration demand in Brazil? *Restoration Ecology*, 25, 509–515.

Supradist, N. 2004. *Economic leakage in tourism sector*. Master of Science in Environmental Management and Policy, Lund University, Sweden.

Waroux, Y. L. P. D. & Lambin, E. F. 2013. Niche commodities and rural poverty alleviation: Contextualizing the contribution of Argan Oil to rural livelihoods in Morocco. *Annals of the Association of American Geographers*, 103, 589–607.

Wry, T. & York, J. G. 2017. An identity-based approach to social enterprise. *Academy of Management Review*, 42, 437–460.

Yergeau, M.-E., Boccanfuso, D. & Goyette, J. 2017. Linking conservation and welfare: A theoretical model with application to Nepal. *Journal of Environmental Economics and Management*, 85, 95–109.

3 Education and skills development

Introduction

One of the cornerstones of the Smart Village concept is social sustainability. This implies a fair and equal society where everyone has a good quality of life. There is overwhelming evidence in the literature that people educated during their childhood lead a better quality life. The concept of the "poverty trap" comes from the fact that the children of poor, uneducated parents often grow up to become poor, uneducated adults who cannot afford to send their children to school and so the cycle continues (Behrman et al., 2017). Barham et al. (1995) showed that children who receive a partial education, although they will obtain a better job than a person with no education, will still not be earning enough to provide a full education for their children, and thus break them out of the poverty trap. The only way to truly break the trap is if the child can receive a full education. These researchers also found that investment in a child's education (human capital investment) is more than compensated by their increased productivity during their working life. This result has been echoed by various other researchers, for example Glewwe and Jacoby (2004) who found that increased human capital investment resulted in better economic growth, while the Center for Global Development (2002) reports that no country has achieved continuous and rapid economic growth without at least 40 per cent of adults being able to read and write. Therefore, investment in education is not only beneficial to the direct recipients but also to society as a whole. It leads to economic growth, lower crime rates and a better quality of life for all.

The benefits of investment in education are widely acknowledged, and the concept has been incorporated into many global campaigns aiming to alleviate poverty and reduce inequality: Article 26 of The Universal Declaration of Human Rights (United Nations, 1948) states "every human being has a right to an education, at least at the elementary and other basic levels, that is free and compulsory"; the second objective of the World Declaration on Education For All states "by no later than 2015 all children, in particular girls, those in difficult circumstances, and those belonging to ethnic minorities, must have access to and must complete government-sponsored, mandatory primary education of good quality" (UNESCO, 1990); and the second objective of the Millennium Development Goals states "ensure that, by 2015, children everywhere, boys and girls alike, will be able to complete a full course of primary schooling" (United Nations, 2015a).

Now that we are beyond the 2015 goalpost of the Millennium Development Goals, it's time to consider whether these objectives are affecting real change for rural communities, and if not why not. Although the goal saw a significant increase in primary school access from 83 per cent in 2000 to 91 per cent in 2015, there are still 54 million children out of school. The impact was better in urban areas, where the proportion of children out of school is 8 per cent, in rural areas the number is double this at 16 per cent (United Nations, 2015b). Thus, rural communities are still a long way behind urban areas. The Millennium Development Goal only intended to provide universal primary education for a single generation, working on the assumption that this will provide a subsequent generation of equality, relieving the need for further intervention. However, it has been argued that the provision of education for all simply moves the lowest common denominator from children with no access to education, to children with access to only basic education (Behrman et al., 2017). The divide between rich and poor, urban and rural will still exist. We need to envision an education concept that goes beyond providing the basics, to ensure we are breaking the poverty trap, bridging the urban-rural divide, and thus giving the next generation of adults a real chance at leading a good life.

The fourth Sustainable Development Goal on education, adopted in 2015, goes beyond the previous Millennium Development Goal to "ensure inclusive and quality education for all and promote lifelong learning" (United Nations, 2017). The specific targets include access to free pre-primary and secondary education, as well as primary. While this should theoretically achieve the goals of poverty alleviation and increased equality, there are challenges beyond the financial cost of education provision that need to be overcome. In this chapter, we will explore some of the unique challenges involved in providing education to rural communities, compared to urban areas.

Firstly, we will consider the provision of a classroom environment to children in sparsely populated rural areas and we will present case studies showing creative solutions to overcoming these issues. Another reality existing in rural communities is multigraded education. This is the situation where only one teacher is available to educate children of different ages and with different levels of knowledge. School curriculums are generally set at the state or federal level and focus on a mono-graded system (i.e. children of the same age are in the same class and are expected to attain the same knowledge). There are many reasons why this system does not fit in a rural setting. We will discuss these reasons and look at cases where a different approach that caters specifically to the multigrade setting can improve educational outcomes for children in these schools.

Providing a place for children to attend schools is only the first step to ensuring they receive an adequate education. Ensuring they attend school is the second key step. Repetition of grades and dropout rates among children from poverty-stricken rural areas are higher than average. In some instances, these can be even worse among ethnic minorities and girls. The reasons behind this will be addressed, and interventions that ensure all children continue to attend school through at least primary school level will be discussed.

Among children who attend school regularly, there are still wide variations in their academic achievement, even at the basic primary level. These barriers to a child's ability to attain education have been linked to their early childhood development (ECD). Providing a child with a nurturing environment that supports their cognitive development during early childhood is thus important, to ensure they succeed at school. The Sustainable Development Goal acknowledges this by setting a target to provide pre-primary education for all. The third section of this chapter will explore options for improving ECD.

In the modern world, the basic tools of education have moved beyond basic literacy and numeracy to include technological literacy. Lack of ICT abilities is strongly linked to the urban-rural divide. Due to the high turnover of ICT equipment in developed economies, there are thousands of second-hand computers available annually, therefore the financial burden of purchasing ICT equipment is somewhat mitigated. A more significant barrier is finding teaching staff with highly developed ICT skills. The other is the way children are taught. In this section we will look at how to not just teach children the technicalities of how computers work, but to engage with technology to empower themselves.

Provision of schooling for children is a long-term strategy aiming to create socially sustainable Smart Villages in subsequent generations. In the interim the current generation of adults needs to be empowered as well. In some part this is an essential component of ensuring positive educational outcomes for their children. Parents need greater awareness of how to nurture their children's cognitive development, especially during the pre-primary years. They also need a good understanding of health and hygiene to ensure their children thrive physically as well as intellectually. Further to this, there needs to be increased awareness among rural communities of sustainable agricultural practices, and other environmental issues to help communities thrive during extreme weather events etc. Thus, this chapter will also explore examples of adult education programmes.

This chapter will conclude by bringing together aspects of the successful education programmes discussed providing recommendations for best practice.

Providing a classroom experience

The typical format of a school that prevails in many developing economies is a series of rooms with individual desks in rows. There is a separate room for each class, and each class contains children of the same age learning the same material. This format, which is rooted in the traditional Victorian era schooling system, does not translate into an effective learning environment for today's rural community for many reasons.

Firstly, recent research has found that the arrangement of space in these types of buildings is not conducive to learning. Uduku (2015) explains that a child's capacity to learn increases if they feel comfortable. Other research has also shown that participating in group learning activities gives children greater confidence and allows them to develop leadership skills (Forero-Pineda et al., 2006). Therefore, having a classroom, with different areas for different activities, that encourages group learning increases a child's ability to attain education.

Tasci (2015) goes further to suggest that the school buildings and ground themselves should provide an educational tool for children to interact with. Especially when it comes to teaching children about the concepts of sustainability, learning within a sustainably designed building will provide the children with a living example of the concepts allowing them to develop a deeper understanding of the issues. Similarly, the landscape design of the school grounds can provide an active learning tool for physics, geometry, botany and ecology as illustrated in Figure 3.1.

Secondly, in many regions, an outside entity, either a representative from the government, or an NGO will be brought in to build the school. Once the builders have gone, there is no one left in the area, with the ability to maintain school buildings and their associated infrastructure. Uduku (2015) gives an example of a charitable organisation building a school for a rural community with solar panels for electricity. However, after a while the solar panels failed, and no one in the region understood how they worked, and thus could not repair them.

Another barrier to providing universal education, particularly in sparsely populated remote areas, is that the school may only be in close proximity to a handful of children, meaning many must travel long distances to attend. This situation may take up a lot of time for some children or be virtually impossible in areas with lack of transport infrastructure.

One option is to provide boarding facilities for children travelling from further away, so they only need to return home at weekends (see Case Study 3.1). While this may be viable in some areas, in many rural areas, families rely on their children to perform household duties, or assist in the family business, particularly where the family runs an agricultural smallholding. Boarding would not be a

Figure 3.1 Children using an adventure playground

viable option for these families, and parents would be discouraged from sending their children to school.

Thus, this emphasis on providing a dedicated physical facility is in some ways misguided. In a smaller community it makes more sense to provide a multipurpose building that can be used as a school during the day then provide other community services during the evening, for example community support group meetings, town council meetings, or adult education classes. The school can potentially

Case Study 3.1 Druk White Lotus School in Ladakh, Northern India (Turner, 2015 and Druk Padma Karpo Educational Society, 2016)

The concept for this school in the remote region of the Himalayas epitomises the ideals of sustainability incorporating the environmental, social and economic, as illustrated in Figure 3.2.

All the building materials are locally sourced, and construction is being conducted using traditional techniques, without the use of machinery. Thus, the local community are able to perform any repair and maintenance work on the building, including replacing any building materials, at low cost and without the need to rely on external help. The consequence of using such traditional techniques is that the building process is very time-consuming. Further to this, the extreme weather conditions experienced in the region mean construction can only occur for three months of the year. Thus each year the aim is to construct a single building, allowing the school to expand over time. The first year of operation, saw only a single class of children at the youngest age group. As these children grow and move up to the next class, they will move into a new building, allowing a new group of students at the younger year level to start.

The buildings are designed to optimise passive heating and cooling, and incorporate solar panels to provide electricity.

The design of the buildings and surrounding gardens has been taken from the Buddhist principles that the majority of the local population believes in. However, the school is secular, welcoming children from all faiths.

The school provides boarding facilities for children travelling from further away to attend.

Outside of school hours the school buildings are available for use by the community, for other purposes including community meetings, and adult education courses. Eventually, the school also aims to incorporate healthcare facilities, not just for students but also for the whole community.

charge a fee for other people using the buildings, providing an opportunity to generate revenue to purchase books and other school equipment. Inversely, an existing building that is not used during the day could potentially be repurposed. For example, in Latin America existing buildings such as community centres, or in some cases even a residential home have successfully been used as a school as part of the *Nueva Escuelas* initiative (Forero-Pineda et al., 2006) (see Case Study 3.2).

Finding teachers to work in remote areas is also a challenge. Teachers are higher educated people, who may not want to live in a remote area, where their own children will be deprived of the superior educational opportunities available in urban areas. Offering extra incentives to teachers to work in a remote area is one option. In developed economies there are examples of recent graduates being

Case Study 3.2 *Nueva Escuelas*, Colombia (Forero-Pineda et al., 2006)

The *Nueva Escuelas* movement was founded in 1976 in Colombia as a new way to bring education to rural communities.

It is a method of multigrade schooling that uses self-directed learning guides to allow children to learn at their own pace. One of the aspects of the initiative is greater involvement of parents and the wider community in the education process.

Evaluation and levelling are flexible, allowing children to meet learning goals in their own time. This not only caters to children of different abilities, but also allows time for children who are required to participate more actively in the family business to achieve the same level of education as other children.

Due to the long-term implementation of the *Nueva Escuelas* programme there have been many assessments of its impact on children, their families and the wider community, in comparison with conventional education programmes. All of these assessments found children from the *Nueva Escuelas* programme outperformed children from conventional schools in literacy and numeracy ability. In fact, children of poor families participating the *Nueva Escuelas* programme, outperformed children of wealthy families, who attended conventional schools. Perhaps more importantly, children from the *Nueva Escuelas* programme were more involved in their community and had higher self-esteem.

One assessment study (Pineda et al., 2006), went further to assess the impact of *Nueva Escuelas* on the families of the children and also on how alumni of the programme were getting on.

The assessment of the parents of these children found some important results. The parents' relationship with their children improved due to the programme, as it helped them develop better communication with their children. These parents also felt the school had an impact on the wider community, promoting a more peaceful society. The assessment of alumni found some very important implications. They were more likely to participate in volunteer work, and also participated in a wider variety of volunteer organisations than alumni from conventional schools. They were more disposed towards participatory democracy that representative democracy. For example, while alumni from conventional schools were happy to participate in voting for community development, alumni from *Nueva Escuelas* were more likely to propose new initiatives and collaborate with their local government. Alumni from *Nueva Escuelas* were also more open to listening to ideas from women and people of different races compared to alumni of conventional schools.

accelerated through a teaching course, with heavily subsidised or in some cases, no fees, on the provision that they teach in a poor performing school for at least one year after graduation (Personal Communication, 2006). However, Banerjee et al. (2017) found that students whose teachers had low job satisfaction did not perform as well academically. Similarly, if teachers are only going to work there for one year, it will be difficult to have a continuous standard of education at the school over time. Thus, forcing teachers to work in a remote location could have negative consequences for the students.

The smaller population in rural areas means there is often too few children to make a full class for each age group. Combined with the problems associated with

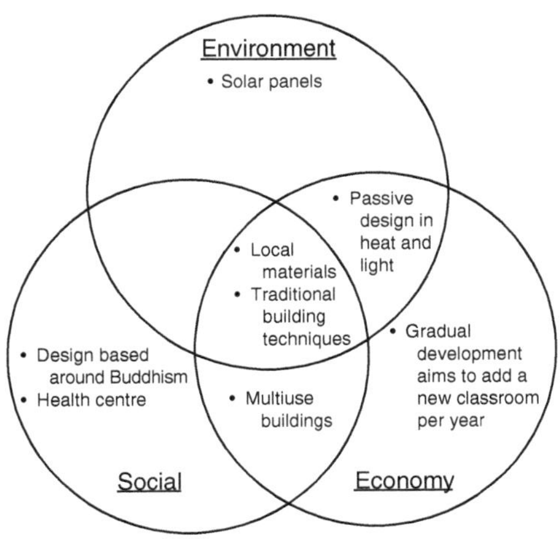

Figure 3.2 The relationship between social, environmental and economic sustainability achieved through the design of the Druk White Lotus School, India

finding teachers discussed earlier, it is difficult or impossible to provide a separate teacher at each class level. Therefore, a single teacher will often be required to teach students of different class levels and ages simultaneously. This situation is known as multigrade schooling, as opposed to conventional monograde schooling systems that prevail in urban areas. In most countries, teachers are trained to work within the monograde system. On top of this, school curriculums are developed at the state or nationwide level, and thus cater to the more common monograded schooling system. This leads to teachers working in a multigrade classroom feeling overwhelmed. Due to the necessity of a single teacher to prepare lessons for multiple classes, compared with teachers in monograded schools who are only required to prepare a single lesson plan, multigrade teachers can become demotivated, which negatively affects child learning (Little, 2006). Many multigrade teachers believe that the children are receiving a lower standard of education than their monograded counterparts.

However, multigrade education systems need not be a disadvantage. In fact, historically teaching students at multiple levels simultaneously was normal. The concept of monograded classes actually emerged fairly recently. To improve the quality of learning in a multigrade environment establishing a separate curriculum with corresponding training for teachers in a multigrade system could provide a solution. Although this seems like it could be difficult and time-consuming, in fact there are many successfully implemented multigrade curricula, which then simply need to be adapted. Little (2006) provides a detailed account of multigrade teaching systems that exist throughout the world. Another advantage of establishing separate educational criteria for rural communities is the freedom to tailor classes to focus more on teaching skills that are relevant to rural life, such as agriculture. This can provide an opportunity to help pervade sustainable agricultural solutions throughout the whole community. For example, in interviews with parents of children that were taught through the *Nueva Escuela* programme, the parents agreed that the way they communicated with their children had improved due to the programme (Forero-Pineda et al., 2006). Similarly, in Australia a campaign to promote greater awareness of the dangers of skin cancer due to excessive exposure to the sun was marketed through schools. The programme known as "SunSmart" was found to be successful at not only improving awareness among children but also among their parents and the wider community (SunSmart, 2008).

The final option we consider is the provision of learning via telecommunications. This is especially a good consideration for areas where single families may be living hundreds of miles from their nearest neighbours. This may invoke images of a virtual classroom experience provided via the internet, an option that may seem out of reach to many communities that do not have access to electricity, let alone internet connectivity. However, in Australia the School of the Air (see Case Study 3.3) has been successfully operating since 1948 (Australian Government, 2016) long before the internet was a thing. The programme initially involved the use of pedal powered radio transmitters, which could still provide a connection to a classroom experience for students living in areas with little or no access to electricity. Of course, in modern Australia, the School of the Air is expanding to incorporate internet technology, leading it closer to a virtual classroom experience. However, its beginnings show that in remote areas with lack of electricity and internet connectivity, it's still possible to provide a remote learning experience.

Case Study 3.3 The School of the Air (Australian Government, 2016)

Australia is 7.7 million square km, with a population of only 24 million people. 85 per cent of those live within 50 km of coastal areas, meaning there is a population of 3.6 million people, in an area of 7.1 million square km. Many of these are located in remote areas, far from any towns. Up until the 1940s, children in these areas either had to attend boarding schools, missing out on spending time with their families, and being away from family farms, where they could be helping out. Or else they would participate via mail, which was also difficult because sending and receiving homework in this way took a long time and meant they often fell behind other students.

In 1946 Miss Adelaide Miethke, a former school inspector who took up a post with the Royal Flying Doctor Service, noticed that most children in these remote locations learnt how to use the radio so that they could call for help if something should happen to their parents. She hit up on the idea that the radio could be used to connect children with teachers and other students and thus the School of the Air was born. It is still operating today, although of course it has moved forward using more modern technology to bring the virtual classroom experience even closer to the students. Each year the school has a camp and sports day allowing these children a chance to meet up with others in person and develop their social skills. These events are always attended by almost all of the students. Thus the programme has been a huge success and could provide an excellent example of how to overcome some of the challenges involved in delivering education to remote communities.

Maintaining attendance

In the previous section we touched on the fact that many children have difficulty accessing their nearest school due to lack of transport infrastructure. There are other barriers to school attendance in rural communities. Many families rely on their children for income either directly, children are working on the family farm or participating in other income generating activities, or indirectly, children are performing household duties, looking after siblings, and thus allowing their parents more time to work. By sending their children to school, these families will lose part of their income, and may not be able to afford basic provisions. Parents in this situation may find it hard to see the value in sending their children to school, as their immediate concern is providing for their family. An example of this is the "Queen of Katwe" (Crothers, 2012). Her mother sent her and all her siblings to work in the market all day. At the end of the day they would have only made enough money to buy food for dinner that night, their only meal of the day. From her point of view, sending her children to school meant no meal.

Thus, in many regions providing a free meal during the school day has been employed to incentivise parents to send their children to school. There are difficulties in implementing a school meal programme. The first issue is deciding whether the meal should be for all children, or only those in need. This will depend upon the level of financial support for the programme, and the variation in the socio-economic situation of the children. When choosing the latter option, deciding on which children qualify to receive the free meal becomes a tricky task. Essuman and Bosumtwi-Sam (2013) found some examples where the implementation had missed the mark and wealthier children were receiving meals, while the poorer ones were missing out. There have also been reported incidents of children being ostracised by their peers for receiving the meal, as it emphasises the fact that these children are poorer. However, there were also reported instances of schools that had a meal for all children in which wealthier children, who had eaten before school, or been given snacks to take to school by their parents would share their free meal with the poorer children. In some instances, poor children were also able to bring home leftover food, for younger siblings.

The timing of the free meal also needs consideration. It is generally provided at lunchtime, several hours after classes have started. For the neediest children, who are arriving at school on an empty stomach, this means that they have to sit in class for several hours before they can eat. This could severely inhibit their ability to engage in class activities and absorb information.

The final issue identified relating to school meals is the quality of the food the children were receiving. Concerns have been raised about whether the meals are thrown together from scraps and leftovers with low nutritional value. There have also been concerns about whether the meals are prepared in a hygienic environment. In many rural schools, it is unlikely that a professionally trained cook would be available to prepare the meals. A positive way to overcome these concerns is to incorporate the meal programme into the school curricula. In South Africa, many schools incorporate gardens where they grow their own food for the school feeding programmes (Uduku, 2015). These gardens provide an interactive learning tool, giving children hands-on experience of agricultural skills that they are most likely to need as adults. This could also provide a good opportunity to teach children about sustainable agricultural practices, which they could pass on to other members of the community, especially their parents. Thus, this could enable the concepts of sustainability to infiltrate the current generation. The children can then also gain hands-on experience of cooking, while also learning about healthy eating and hygienic food handling practices in a fun and interactive way.

The requirement for children to help out at home is not necessarily consistent throughout the year. Families that run a small agricultural business as their main source of income often need their children to participate during harvest time (Kazeem & Musalia, 2017), but the children are free to attend school for the rest of the year. Similarly, Randell and Gray (2016) found that children of agricultural families tend to miss school more frequently during poor weather conditions, especially hot dry summers. They report that this is due to the reduced or failed harvest, requiring the family to chip in to make up for the loss of income. The predictions of climate change analysis indicate that such weather phenomena are set to increase, indirectly depriving children of an education (Randell & Gray, 2016). Thus, the children still have ample opportunity to attend school, but by missing part of the school year, they fall behind the rest of the class, and often need to repeat grades. While this in itself is an inefficient use of already stretched resources in rural schools, it has also been shown that children that repeat grades are more likely to drop out (Kazeem & Musalia, 2017). This is attributed to the social stigma associated with being older than the other children in the year level, and difficulty in being left behind while friends advance through school. This is another example of how the monograded teaching system fails to meet the needs of rural communities. Establishing an education system that emphasises individual learning, rather than arbitrarily assigning educational attainment goals to an age would be a more appropriate approach. This system would also overcome the issues facing the multigrade classes that exist in many rural communities. An individual based system gives children the freedom to learn at their own pace, meaning that after taking time out to help with the family business, they can pick up where they left off. This is much more practical than throwing them back into the class that has advanced in their absence and expecting them to be able to catch up.

An example of a successfully implemented individual based learning curriculum is the *Nueva Escuelas* programme in Colombia (see Case Study 3.2). A key component of this system is providing teachers with specialised training on individualised learning. This gives teachers more confidence in managing a multigrade classroom, while at the same time allows students of different ages and abilities to learn at their own pace in the same space. Children from this system have been shown to have better outcomes than those in "conventional" schooling systems. The *Nueva Escuelas* programme also provides children with learning materials to take home, so they can work in their spare time at home during periods where they are unable to attend school.

Marsh et al. (2016) found that for households that farm animals, vaccinating livestock lead to an increase in school attendance by the children of these families. It is believed that the reduction in loss of livestock, reduced need for antibiotic and acaricide use leads to an increase in income for those families that chose to vaccinate, allowing them to send their children to school. Thus, subsidising farmers seems to indirectly increase educational attainment for children of farming families.

Achieving educational attainment

Although the Millennium Development Goals saw a significant increase in the number of children that have access to primary school education, this has not translated into the same number of young people with basic literacy skills (United Nations, 2017). Thus, although more children are attending school, they are not necessarily attaining education. One reason for this is the different standard of teaching at different primary schools. In India, while most villages have a primary school, secondary schools are mainly found in bigger towns and will take in children coming from many different primary schools. Teachers at the secondary schools were finding that children from some schools had very different levels of knowledge than others, making it difficult for them to set a curriculum that would allow the children who were a bit behind to catch up, without making the other children too bored. The Government of Assam came up with a new initiative to tackle this problem: Teachers from the secondary school were sent to the poor performing primary schools where they would work with the primary teachers, to help them develop their skills so that they could bring the children up to the right level for secondary school entry. The programme is still in a trial phase at a few case study schools but if the results are positive the government plans to expand the programme throughout the state (Government of Assam, Personal Communication).

One of the main reasons attributed to this inability to learn effectively is poor early childhood development. Liu and Hannum (2017) investigated educational attainment of people exposed to poverty at different times during their childhood and found that poverty experienced in early childhood had a more significant negative impact than poverty experienced in later life. Other researchers (Pérez-Escamilla & Moran, 2017, Richter et al., 2017, Lo et al., 2016) claim that provision of early childhood education facilities is equally as important as provisions of a primary school. By the time children reach primary school age, it may already be too late for them to develop their full potential. Therefore, it is important that this is taken into consideration when devising strategies to provide education for all. The Sustainable Development Goal for education incorporates this into its targets by requiring access to pre-primary education as well as primary education.

Early childhood factors that impact educational attainment

To develop the most appropriate pre-primary intervention programme, the specific factors that impact a child's development must be understood. Lack of nutrition is often considered the most important, and thus there are many feeding programmes for infants. It is widely understood that where possible breastfeeding is the best nutritional support for early childhood especially in the first 1000 days. The method of feeding is equally important. There is a widely held misconception that all children should be fed the same amount to a set schedule. However, research shows that all children have different needs and parents should develop the ability to respond to their child's cues of when they are hungry or full to start and stop feeding. Although this comes instinctively to many mothers, some struggle with this and need intervention to help them develop the ability to respond to their child's cues. Breastfeeding is not only a response to hunger, it can also provide a child with comfort in times of distress and should also be used to help keep them calm. Responsive-feeding has been shown to lead to the most positive outcomes for a child's cognitive and social development. Vazir et al. (2013) conducted a randomised controlled trial in India found toddlers of parents that had been taught responsive feeding had improved dietary intake, growth and mental development.

There are also strong correlations between poor childhood development and exposure to infectious diseases (Pérez-Escamilla & Moran, 2017). Teaching parents how to practise good hygiene when feeding their children is therefore another important aspect of early childhood development intervention programmes. Children in poor areas experience higher morbidity rates and disability rates, especially stunted growth, due to poor maternal nutrition, poor nutrition during infancy etc. Children with disabilities are subject to worse circumstances as their parents don't see their value and are less likely to send them to school, or provide them with a stimulating learning environment. It is important to educate parents firstly to have better hygiene practices to reduce incidents of childhood disability, and secondly, where children are already disabled, to have a better understanding of how the child will be limited by the disability, and how they can be nurtured to achieve their full potential despite the disability.

While nutrition and hygiene interventions have been shown to have a strong influence over a child's development, they are not the only factors to consider. Engaging infants in learning stimulation activities, education, responsive parenting and social protection have equally important impacts on the future ability of a child to attain education, and be successful during adulthood (Pérez-Escamilla & Moran, 2017). Lo et al. (2016) state "epigenetic, immunological, physiological, and psychological adaptations to the environment occur from conception, and that these adaptations affect development throughout life". Luby et al. (2013) performed a long-term neuroimaging study on children from varying backgrounds. They looked at the relationship between brain matter volume and socio-economic status, parenting style and exposure to stressful life events. They confirmed through imaging of the physical brain structure, what has previously been shown through psychological and educational tests: that poor socio-economic status correlates to lower volume of brain matter. However, they also found that hippocampal development was more strongly correlated with parenting style. Children raised by nurturing parents had better hippocampal development than those raised by hostile parents, independent of income. Their exploration of exposure to stressful life events found that children with nurturing parents were more able to cope after a stressful life event. These results have important implications to children from poor rural families in developing economies who are often exposed to wars or other stressful situations.

Economics of pre-primary programmes

In the US, analysis of early intervention programmes found that children who participated in the programmes had increased academic achievement and employment, and reduced incidents of poverty, special education placements, delinquency and crime. Economically, the programmes were found to save from 1.80 to 17.07 USD per dollar spent through societal benefits (Scharf et al., 2017). Richter et al. (2017) calculated that the cost of scaling up existing health-based pregnancy and early childhood interventions would be on the scale of between 20 and 70 cents (USD) per person per year (the variation was due to the modelled level of coverage of the intervention). Conversely, the costs of inaction, although dependent on many country specific factors, generally tend to outweigh the costs of intervention. For example, in India where stunted growth of children living in poverty is prevalent, the cost to the government for not intervening is around 2 billion USD. This takes into consideration the current cost to the healthcare system for stunted growth minus the costs of the intervention programme.

Access to early childhood education has been shown to be related to the urban-rural divide. Nonoyama-Tarumi and Ota (2010) studied early childhood education in 24 countries and found that twice as many urban children had access compared to rural children. Children from wealthier families were also twice as likely to attend early childhood education programmes.

Richter et al. (2017) propose that ECD programmes need not be developed from scratch but can simply build on existing programmes. As an example, the WHO-UNICEF Care for Child Development (CCD) programme started out as integrated management for childhood illnesses, implemented by a combination of home visitors or at facilities. The programme was scaled up to improve early childhood nurturing, through counselling of parents during a visit. The worker could share knowledge about nurturing parenting techniques and good hygiene practices, while tending to the sick child. Therefore, no extra time during the visit was required, and the only economic scale-up was training the existing staff in the counselling techniques. The results of the programme showed a successful improvement in the living conditions of the children, and a corresponding improvement in their cognitive development.

In developing economies parents are often required to work long hours, just to make ends meet, and therefore do not have time to spend caring for their children. In many instances, the role of caring may be passed on to an older sibling, especially female. Therefore, providing education to the older siblings on infant nurturing may be essential. This could have the bonus of better preparing the child for parenting when she has her own children.

ICT and libraries

In the modern world, it is essential to have information and communication technology skills to obtain employment. Even for people wishing to start their own business, a lack of internet skills can hinder their success as more people turn to the internet to find information about local businesses, and do their shopping etc. Thus, it is essential that any education programme includes ICT skills development. Schools in many regions have been provided with ICT equipment through charitable donations; however, this does not always translate into children attaining ICT skills (Rubagiza et al., 2011). In some areas, the teachers did not know how to use the computers, and therefore were unable to pass the knowledge onto the students. Therefore, it is essential the ICT training for teachers is provided along with the provision of equipment. In other instances, ICT skills were taught from books, rather than through interaction with the equipment. Again, this did not lead to effective attainment of ICT skills.

This is an issue relating to the urban-rural divide. In many urban areas, there are many internet cafes, and children will spend their spare time here, learning how to engage with technology through social interaction. Such facilities are far less common in rural areas, thus children in rural communities are relying solely on the school to provide them with ICT access (Rubagiza et al., 2011).

Rubagiza et al. (2011) and Rowsell et al. (2017) both argue that to really develop ICT skills, equipment should be used to teach different subjects, rather than just teaching ICT skills. Children develop ICT skills far more effectively in this way. One way to overcome the issue of not having teachers with relevant ICT skills available is to develop self-teaching software and provide this to schools along with the ICT equipment. While this sounds good in theory, it is important that the self-teaching software is developed with the customs of the local community in mind. For example, a self-teaching computer programme on health and hygiene, developed for South Africa, was passed onto schools in Rwanda (Rubagiza et al., 2011). Due to the more conservative values of Rwandan culture the teachers were offended by the content of the programme and refused to allow the children access to it.

The type of ICT equipment provided can also influence the attainment of ICT literacy. Learning ICT skills through interaction with smartphones, shows lower academic achievement than through using computers (Rowsell et al., 2017).

Traditional teaching materials are also essential for children's learning in the multigrade rural classroom. Having access to library facilities, which children can use without the aid of the teacher, allows children the chance to engage in independent learning activities, allowing the teacher to work with other children.

Adult education

In the previous section we discussed the importance of providing parents with knowledge of nurturing parenting practices, to ensure the best outcomes for the children. It has also been shown that children with higher educated parents, especially the mother, achieve greater academic success.

Teaching basic literacy and numeracy skills to adults is generally more challenging than teaching to children. In India, a computer-based programme for teaching adult literacy, mainly aimed at women, was tested (Deshpande et al., 2017). It was found that people who had no skills prior to taking the course did not improve their skills significantly. However, for people who had even only minor abilities, such as recognising a few letters, showed significant improvement through the programme. It may be more appropriate to emphasise skills learning, rather than literacy and numeracy: teaching agricultural skills in the context of sustainability, entrepreneurship – teaching local people about ways to use natural areas to generate income, without destroying them, other income generating opportunities, hygiene and other health issues.

It is important to give adults a range of skills if they are to become truly self-sufficient. For example, teaching tailoring skills or horticulture, aquaculture, beekeeping etc. is a great first step in giving them the means to earn an income. However, to truly establish these skills as a business they will also need to have basic marketing skills to ensure they are able to sell all that they produce. They will also need to have banking and money management skills to manage their income and use it to continue to improve their quality of life. Therefore, it is essential that business and money management skills are taught alongside any trade skills, so people can become truly self-sufficient. It's also good to encourage community members to develop businesses together, rather than individual families, to promote greater income generation (Banerji, 2015). This could also make it easier to set up funding for community needs such as school buildings, public toilets etc.

Sustainable agriculture is also an essential component of improving quality of life in rural areas, where agricultural activities are the main source of income for most people. As we will see in Chapter 4, sustainable agricultural practices can be complex and will require training for many of the uneducated workers, currently engaged in this field. This educational approach has been shown to be effective in Bangladesh where information about the use of pesticides was distributed amongst farmers. The farmers that participated in the course reduced their overall use of pesticides, and were more likely to use them following strict safety guidelines (Gautam et al., 2017). In this instance the benefits of farmer education are multiple, the farmer is protecting their own health, as well as the health of consumers of their products, and by reducing the amount of synthetic pesticides released into the environment, they are protecting biodiversity.

Another initiative to help farming communities improve their quality of life is the *Solidarité Rurale* in Porto-Novo, Benin (Dara, 2016). The education programme was founded in 1988 by Salim Dara to improve the quality of life of young adults by providing them with a hands-on agricultural learning experience. Since 2007 he altered the programme to emphasise environmentally friendly farming practices. Although other agricultural training programmes exist throughout Africa the advantage of the *Solidarité Rurale* is its emphasis on small scale farming. Many other training programmes work within large scale industrial farming, which would have prohibitively expensive start-up costs, especially for an inexperienced farmer. At *Solidarité Rurale*, the course teaches how to grow enough food to generate a modest income, within a typical plot of land owned by a single family. The programme has inspired changes in other areas for example, the agricultural curricula at universities are shifting so that students gain technical and field expertise prior to completing their certification studies, developing the practical skills employers seek and the experiences they need to start their own successful farming ventures. More than 600 people visit the centre each year to learn from this model.

Reducing conflict through education

While the Millennium Development Goals saw the number of children attending primary school increase significantly, in conflict affected areas, the reverse is true. Between 1999 and 2012 the number of children out-of-school increased from 30 per cent to 36 per cent, particularly in North Africa and Southern Asia (MDG Monitor, 2017). The evaluation of the Millennium Development Goals also noted that approximately 50 per cent of children out of school were from conflict affected areas. The Sustainable Development Goal notes that education is crucial to fostering tolerance between people and thus contributing to a more peaceful society (United Nations, 2017). This leads us to wonder is there a pattern of lack of education in conflict affected areas leading to children becoming embroiled in the conflict as adults? Could the provision of education to the new generation help foster a more harmonious environment by teaching tolerance?

At the Actonville Primary School in South Africa primary students regularly create and oversee "community heart and soul" days (Maphatsoe, 2017). Similar programmes are being established all across Africa as initiatives that allow students to apply what they learn in the classroom to real world situations. When these children leave school, they are better prepared for work, as they have experience of leadership and working with a team. Children are also required to prepare a budget as part of these experiences, giving them greater understanding of finance as well.

The *Nueva Escuela* programme has also incorporated elements to promote peaceful relationships between students (Forero-Pineda et al., 2006). Colombia, where the programme was created, has the highest homicide rates in the world.

Recommendations for best practice

As with all the sections of this book there is no single recommendation for best practice as each situation is unique.

It is strongly recommended that schooling begins earlier than primary school age. This does not necessarily need to take the form of providing a school room experience, intervention with parents to ensure they are aiding their children's cognitive development during the formative early years of life should be sufficient.

In establishing primary schooling, the first step is to assess whether there are enough students to justify a dedicated school building and whether teachers are available locally to staff the school, or whether a virtual classroom experience could be more appropriate.

Where a building is required the next step is to determine whether there is an existing building available during school hours, with appropriate facilities (sanitation, water etc.) to enable the students a safe and healthy learning environment. The building should also be designed to maximise the learning experience by providing opportunities for hands-on and interactive learning. This could be especially important considering the growing need for awareness of sustainability, as having a school building with sustainable features that they can see and interact with will make a stronger impression on children, than simply telling them about it.

If it is determined that a new building is essential and economically justified, it is important that the building is designed to fit in with the cultural values of the community and where possible adheres to the latest research on building design for effective learning.

Generally, for the remote rural village provision, evidence supports the idea of a providing a free meal to the children to incentivise both the children and their parents. As agriculture is the predominant income generating activity in these regions, incorporating an interactive agriculture experience as part of the learning process is recommended, with the additional benefit of allowing the children to provide the food for their own lunches, teaching them sustainable agricultural practices to carry on and share with their families and other members of the village. Incorporating food preparation as part of the school curriculum would also be an effective way to teach children safe food handling and preparation, potentially preventing many of the childhood diseases that pervade poor communities.

It is also recommended that the meal be served in the earlier part of the day so that children who were unable to eat at home, are not disadvantaged during class time, where hunger has been shown to inhibit their ability to learn.

References

Australian Government 2016. *The school of the air and remote learning* [Online]. Canberra. Available: www.assoa.nt.edu.au/visitors-centre/the-centre/history [Accessed 15 March 2018].

Banerjee, N., Stearns, E., Moller, S. & Mickelson, R. A. 2017. Teacher job satisfaction and student achievement: The roles of teacher professional community and teacher collaboration in schools. *American Journal of Education*, 123, 203–241.

Banerji, R. (Director) 2015. *Voices from the forest*. India.

Barham, V., Boadway, R., Marchand, M. & Pestieau, P. 1995. Education and the poverty trap. *European Economic Review*, 39, 1257–1275.

Behrman, J. R., Schott, W., Mani, S., Crookston, B. T., Dearden, K., Duc, L. T., Fernald, L. C. H. & Stein, A. D. 2017. Intergenerational transmission of poverty and inequality: Parental resources and schooling attainment and children's human capital in Ethiopia, India, Peru, and Vietnam. *Economic Development and Social Change*, 64(4), 657–697.

Center for Global Development 2002. Rich world, poor world: A guide to global development. Washington DC: Center for Global Development.

Crothers, T. 2012. *The Queen of Katwe: One girl's triumphant path to becoming a chess champion*. New York: Scribner.

Dara, S. 2016. *Solidarite Rurale: Transforming agro-education; promoting small farms as gateway for self-employment* [Online]. Available: www.changemakers.com/node306408/entries/solidarit%C3%A9-rurale [Accessed 16 January 2018].

Deshpande, A., Desrochers, A., Ksoll, C. & Shonchoy, A. S. 2017. The impact of a computer-based adult literacy program on literacy and numeracy: Evidence from India. *World Development*, 96, 451–473.

Druk Padma Karpo Educational Society 2016. Druk Padma Karpo School. Available online www.dwls.org/. [Accessed 2 October 2018].

Essuman, A. & Bosumtwi-Sam, C. 2013. School feeding and educational access in rural Ghana: Is poor targeting and delivery limiting impact? *International Journal of Educational Development*, 33, 253–262.

Forero-Pineda, C., Escobar-Rodrìguez, D. & Molina, D. 2006. Escuela Nueva's impact on the peaceful social interaction of children in Colombia. *In*: Little, A. W. (ed.) *Education for all and multigrade teaching*. Dordrecht, The Netherlands: Springer.

Gautam, S., Schreinemachers, P., Uddin, M. N. & Srinivasan, R. 2017. Impact of training vegetable farmers in Bangladesh in integrated pest management (IPM). *Crop Protection*, 102, 161–169.

Glewwe, P. & Jacoby, H. G. 2004. Economic growth and the demand for education: Is there a wealth effect? *Journal of Development Economics*, 74, 33–51.

Kazeem, A. & Musalia, J. M. 2017. The implications of ethnicity, gender, urban-rural residence, and socioeconomic status for progress through school among children in Nigeria. *Social Indicators Research*, 132, 861–884.

Little, A. 2006. *Education for all and multigrade teaching: Challenges and opportunities*. Dordrecht, The Netherlands: Springer.

Liu, X. & Hannum, E. 2017. Early poverty exposure predicts young adult educational outcomes in China. *China Economic Review*, 44, 79–97.

Lo, S., Das, P. & Horton, R. 2016. Early childhood development: The foundation of sustainable development. *Lancet*, 389, 9–11.

Luby, J., Belden, A., Botteron, K., Marrus, N., Harms, M. P., Babb, C., Nishino, T. & Barch, D. 2013. The effects of poverty on childhood brain development: The mediating effect of caregiving and stressful life events. *JAMA Pediatrics*, 167, 1135–1142.

Maphatsoe, K. 2017. Community is uplifted in Actonville. *Benoni City Times*, 23 October.

Marsh, T. L., Yoder, J., Deboch, T., McElwain, T. F. & Palmer, G. H. 2016. Livestock vaccinations translate into increased human capital and school attendance by girls. *Science Advances*, 2.

MDG Monitor. 2017. *MDG 2: Achieve universal primary education* [Online]. United Nations. Available: www.mdgmonitor.org/mdg-2-achieve-universal-primary-education/ [Accessed 19 July 2018].

Nonoyama-Tarumi, Y. & Ota, Y. 2010. Early childhood development in developing countries: Pre-primary education, parenting, and health care. *In*: UNESCO (ed.) *Paper commissioned for the EFA Global Monitoring Report 2011, The hidden crisis: Armed conflict and education*. New York: UNESCO.

Pérez-Escamilla, R. & Moran, V. H. 2017. The role of nutrition in integrated early child development in the 21st century: Contribution from the Maternal and Child Nutrition Journal. *Maternal and Child Nutrition*, 13.

Randell, H. & Gray, C. 2016. Climate variability and educational attainment: Evidence from rural Ethiopia. *Global Environmental Change*, 41, 11–123.

Richter, L. M., Daelmans, B., Lombardi, J., Heymann, J., Boo, F. L., Behrman, J. R., Lu, C., Lucas, J. E., Pérez-Escamilla, R., Dua, T., Bhutta, Z. A., Stenberg, K., Gertler, P. & Darmstadt, G. L. 2017. Investing in the foundation of sustainable development: Pathways to scale up for early childhood development. *Lancet*, 389, 103–118.

Rowsell, J., Morrell, E. & Alvermann, D. E. 2017. Confronting the digital divide: Debunking brave new world discourses. *The Reading Teacher*, 71, 157–165.

Rubagiza, J., Were, E. & Sutherland, R. 2011. Introducing ICT into schools in Rwanda: Educational challenges and opportunities. *International Journal of Educational Development*, 31, 37–43.

Scharf, R. J., Maphula, A., Pullen, P. C., Shrestha, R., Matherne, G. P., Roshan, R. & Koshy, B. 2017. Global Disability: Empowering children of all abilities. *Pediatrics Clinics of North America*, 64, 769–784.

SunSmart 2008. Skin cancer prevention: A blue chip investment in Victoria. Victoria: SunSmart Cancer Council.

Tasci, B. G. 2015. "Sustainability" education by sustainable school design. *Procedia – Social and Behavioral Sciences*, 186, 868–873.

Turner, T. (2015) Druk dragon garden in Ladakh: The design and concept in 2015. Presentation to Arup Associates. Available: www.youtube.com/watch?v=3l4Cosg6eqs [Accessed 2 October 2018].

Uduku, O. 2015. Designing schools for quality: An international, case study-based review. *International Journal of Educational Development*, 44, 56–64.

UNESCO 1990. World declaration on education for all. Jomtien, Thailand.

United Nations 1948. The universal declaration of human rights. Paris.

United Nations 2015a. *Millennium development goals and beyond 2015* [Online]. United Nations. Available: www.un.org/millenniumgoals/bkgd.shtml [Accessed 21 December 2017].

United Nations 2015b. The millennium development goals report. New York.

United Nations 2017. *Sustainable development goals 4: Quality Education* [Online]. United Nations. Available: www.un.org/sustainabledevelopment/education/ [Accessed 15 March 2018].

Vazir, S., Engle, P., Balakrishna, N., Griffiths, P. L., Johnson, S. L., Creed-Kanashiro, H. et al. 2013. Cluster-randomized trial on complementary and responsive feeding education to caregivers found improved dietary intake, growth and development among rural Indian toddlers. *Maternal and Child Nutrition*, 9, 99–117.

4 Agriculture

Introduction

At the heart of the UN's sustainable development strategy is the first Sustainable Development Goal of "No Poverty". Hunger and malnutrition are manifestations of poverty, which is addressed by the second goal of "Zero Hunger" and increased food security. As the rate of global population increases the challenge of providing secure sources of food for people, coupled with a host of environmental problems currently facing the planet, including soil depletion and increasing aridity in some parts of the world, food security is a primary concern. The proportion of hungry people in the world significantly decreased, from 18.6 per cent in 1990–1992 to 10.9 per cent in 2014–2016, despite an ever-increasing world population. However, there are still 795 million people suffering from chronic undernourishment (Hunger Notes, 2016). It has been asserted that the problems of world hunger are not due to a lack of food production capability; in 2008, for example, worldwide food production reportedly reached 2790 kcal per person per day (Hunger Notes, 2016). The inability of people to generate sufficient income to purchase food does, however, continue to be a prime driver of a lack of food security and associated malnourishment.

In rural communities agriculture often provides the main source of income; in fact over one billion people worldwide are employed in agriculture, which equates to approximately one in three of all workers (United Nations, 2015). In sub-Saharan Africa and parts of South Asia the share of people working in agriculture increases to more than 60 per cent. Increasing the productivity of smallholder farms in developing economies is seen as key to alleviating hunger and poverty. The UN has recently declared that "if done right, agriculture, forestry and fisheries can provide nutritious food for all and generate decent incomes, while supporting people-centred rural development and protecting the environment" (United Nations, 2015). This chapter focuses on options for increasing farm productivity in a sustainable way as part of a Smart Village development strategy.

Agriculture is often seen as the epitome of the rural idyll, working in the great outdoors, in harmony with nature. In reality, modern intensive agricultural practices most often work against nature, reducing biodiversity, introducing synthetic chemicals into the environment, depleting soils of their nutrients, and destroying natural waterways. Nature also works against farmers with: unpredictable periods of drought or flooding and other natural phenomenon intermittently destroying crops; outbreaks of crop pests; or livestock killed by disease.

Agriculture can be a very difficult and often an unreliable source of income. In developing economies, many farmers are uneducated, and have little understanding of the biological and physical processes underlying food production, instead relying on luck with the weather to provide them with good harvests. As the effects of climate change set in, especially in sub-Saharan Africa and south Asia, many years of hardship due to droughts and/or flooding are predicted for these regions. Many farmers rely on government subsidies to survive, which often encourage the use of synthetic fertilisers and chemical pesticides that might provide temporary relief for a single harvest, but do not lead to long-term sustainable farming solutions.

Defining smart, sustainable agriculture is complicated because it is a combination of two conflicting goals: on the one hand, agriculture should aim, as far as possible, to work in harmony with the environment. The use of synthetic chemicals for fertilisation and pest control should be minimised, livestock should be raised in a humane way, with space to graze and a diet that reflects the way they would eat if left to their own devices. On the other hand, the goal should also be to maximise the amount of food produced per unit land area to minimise the total amount of land converted for agricultural purposes. Hertel and Baldos (2016) note that agricultural expansion is a leading contributor to greenhouse gas emissions, and one of the driving factors in biodiversity loss. Forests, that contain high biodiversity, not to mention their role as carbon sinks, are being converted to agricultural land at an alarming rate. The Food and Agriculture Organization of the United Nations estimates that 33 per cent of the world's land area is currently being used for agricultural purposes (FAO, 2017).

So, which of these goals are a truly smart approach? Can there be a solution that somehow incorporates both goals? Either way, it is clear that agricultural production must continue to intensify in order to feed the ever-growing population and seek to maintain and increase the income generation of smallholder farmworkers. But this intensification must be done in a sustainable way so that the earth's remaining natural resources are enhanced by agriculture, rather than being decimated as has been the case in many developed countries where agricultural intensification and loss of natural forests has resulted in disaster. We need to ensure our agricultural production can provide enough food and income to sustain the current generation while not destroying the chance for future generations to support themselves.

Currently, agricultural efficiency is considered as produce per unit of land area. Struick et al. (2014) propose that broadening the definition of agricultural efficiency from produce per unit land area, to produce per unit of input could be a more sustainable approach. Input would incorporate not just the land but fertilisers, pesticides, water and so on. To illustrate this idea, consider the use of synthetic nitrogen fertilisers. The use of these fertilisers increased cereal production by 50 per cent between 1960 and 1995. While this seems like a significant increase in agricultural efficiency in terms of produce per unit land area, the amount of synthetic fertiliser needed to be used increased 700 per cent. Thus, in relation to fertiliser use the amount of cereal produced was very inefficient. Expanding our definition of efficiency will thus help us get closer to some form of sustainable intensification.

Organic agricultural practices are often more sustainable and are becoming more popular throughout the world. In 2015, 50.9 million hectares, across 179 countries, were cultivated organically (Mie et al., 2017). Organic produce often comes at a premium cost, attributed to the fact that these practices reduce yields. If this is the case then moving towards an entirely organic approach to agriculture may not be considered sustainable, due to the required increase in cultivated land it requires per unit of produce. However, it could be possible to incorporate some organic methods with more conventional agricultural practices to achieve a more ideal sustainable intensification. In this chapter, we compare conventional and organic approaches from different aspects of agriculture to determine methods that come closest to achieving sustainable intensification. First, we look at crop management, discussing options for managing soil moisture and nutrient content, and pest management, both during the growth cycle and post-harvest storage. Then we look at issues relating to livestock management including feed, manure management and disease prevention. The final section discusses options for best practice agriculture in Smart Villages.

Crop management

Modern agriculture mainly consists of a "monoculture" approach, where the same crop is grown on the same plot of land, season after season. Plant selection is based on response to demand, with little consideration to soil and climate suitability to the growing of the specific crop species. Over time, this has seen the diversity of crops grown dwindle. Dwivedi et al. (2017) found that 91 per cent of calories consumed worldwide come from just 30 species of plants. One of the major drives towards the reduction in crop varieties is the desire for emerging economies to emulate the "western" diet, where most calories come from meat or dairy (Khoury et al., 2014), even though it has been demonstrated that a high animal food diet has been linked with many diseases, such as coronary heart disease. The approach is also intensive, emphasising the quantity of the outputs, with little consideration to the quality of the crop in terms of its contribution to the nutritional health of consumers. Such intensive monocultural farming has depleted soils of the nutrients essential to plant growth and increased plant vulnerability to pest invasion. In the following sections, we will discuss in detail the different approaches farmers have taken to overcome these problems, and their impacts on human health and the environment, to determine methods that make them sustainable. Many of the approaches in current use are effective in the short term, but will fail in the long term, thus are not sustainable approaches and we will discuss these also.

Table 4.1 Types of irrigation

Irrigation type	*Description*
Surface	The crop area is flooded
Sprinkler	The crop area is sprinkled with water (in an imitation of rainfall)
Drip	The water is dripped above the root zone
Underground	Pipes are buried in the soil, with outlets corresponding to where the plants are. Water is distributed through these. These can be computerised so that the water is supplied in a more targeted way. For example, a moisture sensor could trigger the system when it falls below a certain threshold
Sub-irrigation	Where groundwater is relatively close to the surface, it can be raised to dampen the root zone

Reference: FAO, 2002

Irrigation

Plants need a reliable supply of water to grow yet on 83 per cent of the world's cultivated land, where 60 per cent of the world's food supply comes from, farmers rely solely on rainfall for water provision (FAO, 2002). This approach is becoming less practical each year as the impacts of climate change continue to increase periods of drought, particularly in developing economies. Introducing irrigation in innovative ways into effected regions may become essential to ensure an adequate food supply for an ever-increasing population. There are five types of irrigation that can be considered, as shown in Table 4.1.

Surface and sprinkler irrigation are currently the most popular techniques; however, they are also the least efficient. A lot of the water is lost during application, through evaporation, and they can lead to waterlogging and salinisation of soils. This would not be a good initiative to introduce into regions with scarce or difficult to access water resources. (See Case Study 4.1 for an example of the negative consequences of inefficient irrigation.)

Case Study 4.1 The Aral Sea (France 24, 2017)

The Aral Sea, located on the border between Kazakhstan and Uzbekistan, was once the fourth largest inland lake in the world. The lake was mainly fed by two rivers, the Amy Darya in the south and the Syr Darya in the east.

During the Soviet era, the rivers were diverted to irrigate crops including fruits and cereals, but mainly cotton. The pipes for the irrigation were poorly laid, and thus the process was very inefficient, leading to as much as 70 per cent of the water being lost during transfer from the rivers to the farms. On top of this, the region has a very arid climate, and cotton is a very water intensive crop. This led to overexertion of these two rivers and the size of the lake reduced by 80 per cent between 1960 and 2000.

The Aral Sea was surrounded by small villages, whose main livelihood revolved around fishing. Aside from the many anglers, thousands more were employed in the canneries, where the fish was packaged and distributed throughout the Soviet Union. These villages are now ghost towns, and most residents had to move away to find alternative employment after the sea dried up.

While the water disappeared from the sea, the amount of salt remained the same, thus salinity in the region began to make the air quality very bad, and the soils less and less fertile. This led to a huge increase in the use of fertilisers and pesticides to maintain output from the agricultural industry. The air quality in the area has become very bad and the health of the remaining residents severely impacted.

In 2005, a project was implemented by the World Bank to restore the sea, however, only on the Kazakhstan side. The Uzbekistan side is still completely dry. In Kazakhstan the lake is now approximately half its original size and the fishing industry and tourism have returned to the area, slowly boosting the economy and allowing children to return to school etc.

Drip and underground irrigation are far better in terms of water efficiency, and more effective at ensuring that water reaches the targeted plant roots.

Makombe et al. (2017) conducted a comparative analysis of two different types of irrigation systems both with each other and with rain-fed systems. The first system was a traditional one where a trench is dug in the earth allowing water to flow through to the crops from a nearby water source. The trench needs to be re-dug each season. The second system is a newer irrigation approach where the trench is concreted allowing it to last multiple seasons. The results of their study found that the concreted channelised irrigation system resulted in the highest productivity. Interestingly, the rain-fed system had better productivity than the traditional irrigation system, suggesting this to be the least preferable option. However, both the concreted channel irrigation and rain-fed systems have some technical inefficiencies and could benefit from efficiency improvements in Smart Villages.

There are now many examples of irrigation for farmlands using effluent water. This has many advantages, aside from providing a reliable water supply for plants, as it can reduce the need to use potable water and divert wastewater from natural waterways where it can cause environmental hazards. Wastewater also contains soil nutrients such as nitrogen and phosphorus, which can be detrimental to the quality of natural waterways, but provide a nutrient benefit to farmlands, as discussed in the following section. The potential downside to effluent water use on crops is the risk of contamination. Wastewater can contain bacteria that are harmful to human health, particularly those that cause gastrointestinal disease. These bacteria are killed at the kind of high temperatures used in everyday cooking, and therefore it is much safer to use on the types of crops that are always cooked before consumption. Using wastewater on products that are consumed fresh could be dangerous unless the water is treated to remove bacteria prior to irrigation. Effluent wastewater can also contain Potentially Toxic Elements (PTEs) such as heavy metals, or other chemicals found in medicines and cosmetics. Some of these elements are known to cause health problems even at relatively low concentrations (Murtaza et al., 2017). These elements can be removed by treating the water prior to irrigation, however, in some areas this practice is not currently required, and soils may already be contaminated. Designing constructed wetlands as biofiltration water treatment is one way to pre-purify water before being used for crop irrigation. The uptake of PTEs into plants varies depending on the plant species. There are plants that will not uptake PTEs at all and identifying these could be one solution to preserving the viability of agricultural activity on soils that are already contaminated.

Soil health

Intensive monoculture farming has depleted soils of the nutrients essential for plant growth (Gardiner & Miller, 2008). The use of fertilisers in some form is now essential to enable farmers to produce crop yields sufficient to provide them with a liveable income. This has led to a massive increase in fertiliser use globally, for example, UNESCO reports that in the state of Assam, India, fertiliser use increased between 1996 and 2003 from 14.2 kg to 46.5 kg per hectare (Mathur et al., 2006). Although this has led to a corresponding massive increase in global food production, there have also been some significant negative side effects. Next, we will look at two of the most significant nutrients required for plant growth, nitrogen and phosphorus, and the different ways they are being replenished through the use of artificial fertilisers, and the impact this is having on the environment and human health.

Nitrogen

The most essential nutrient required for plant growth is nitrogen. To be effectively taken up by plants nitrogen must be in a particular form, otherwise it can be lost to the atmosphere through the microbial processes of nitrification (the transformation of ammonium to nitrate) and denitrification (the transformation of nitrate to other nitrogen gases) (Gardiner & Miller, 2008, White, 1997). These processes can lead to emissions of N_2O, a greenhouse gas with 297 times the radiative forcing of CO_2, and an ozone depleting substance. A nitrogen fertiliser, in the form required for uptake by plants, can be synthetically manufactured. The process for this is combining the hydrogen found in natural gas with atmospheric N_2. A significant amount of energy is required to break the N_2 bond, making this process highly energy consumptive (ADAS, 2009). Thus, the increasing cost of energy, combined with the depletion of natural gas resources is leading to the increase in the price of synthetic nitrogen fertilisers and could see its production become prohibitively expensive for small time farmers in the future (Shafiee & Topal, 2009). In many developing economies, there are already government subsidies for fertiliser costs, as a measure to retain food security (Stamoulis & Lipper, 2012).

Aside from the environmental implications of synthetic nitrogen fertiliser production, there are also significant environmental risks associated with its use. Synthetic nitrogen fertilisers are often distributed on top of the soil, too early in the lifecycle of the plants for uptake. Fertiliser that is not uptaken by plants can be converted to N_2O as described previously. Other nitrogen compounds can also be released to the atmosphere, where they can have negative impacts on human health including cardio-pulmonary ailments (UNESCO & SCOPE, 2007), cancers and liver damage. There are also concerns relating to the presence of nitrogen compounds in drinking water and their potential impacts on health but our understanding of this is still evolving.

Excess nitrogen fertiliser is also transported by wind and rain, and deposited into waterways. Here, it causes eutrophication, manifest as the rapid growth of algal blooms on the surface of waterways. These algae absorb oxygen from the water, and block sunlight from reaching the flora and fauna that live below the surface. Without oxygen and sunlight, underwater biota cannot survive, and thus the waterways become dead zones, devoid of fish and other aquatic species. This is a huge issue for nearby residents who rely on fishing as a source of food and/or income. Because N fertilisers are often applied too early, and partially lost to the environment, farmers often apply far more fertiliser than necessary. It is erroneously considered to be a "more is more" system, where the more fertiliser applied, the greater the crop yield. This situation is sometimes exacerbated by government subsidies, which hide the true costs of excess fertiliser use. Much of the applied fertiliser is being lost to the environment, and crop yields would still be retained if far less fertiliser were applied at the most appropriate time and using methods of application that will best encourage plant uptake. Injecting the fertiliser into the soil, rather than distributing on the surface can also significantly reduce fertiliser use, without any impact on crop yields. Although the labour and/or equipment required for this practice is in many cases prohibitively expensive, especially in regions where the cost of fertiliser is heavily subsidised.

In some instances, timing the application of fertiliser may not be possible due to labour availability. Other, more practical, options include the use of slow release fertilisers or inhibitors (Rurak, 2012). Slow release fertilisers can either be products that naturally degrade slowly in the soil environment, or products that have a physical coating that dissolves slowly. Inhibitors contain bacteria that prevent the processes – such as nitrification and denitrification that transform the fertiliser into other forms of nitrogen that are not suitable for plant uptake – from occurring. Either way, these allow only a portion of the nutrients to be released at a given time, allowing longer for the fertiliser to penetrate into the soil, and reach the plant that it is targeted at and preventing nitrogen release into the wider environment.

Phosphorus

Phosphorus is another essential nutrient needed for plant growth. Although modern agricultural practices have not depleted it from soils, it is no longer found in a form readily available for plant uptake. Phosphorus fertilisers are manufactured from phosphate rock, a scarce resource with no synthetic substitute. It is expected peak production will be reached by 2030 (ADAS, 2009). Phosphate rock also contains the heavy metal Cadmium (Cd), an element which is toxic to human kidneys, can demineralise bones, and is a known carcinogen (Mie et al., 2017). As stocks of phosphorus rock are depleting the concentration of Cd in mineral fertilisers is increasing, leading to concerns over its accumulation in plants, and thus in human diets. Limited research has explored the transfer of Cd from fertiliser application to human food systems. Uranium is also present in phosphate rock. There are concerns it is accumulating in soils from the use of mineral fertiliser and leaching into waterways – a process that is not clearly yet understood (Mie et al., 2017). Similar concerns abound in relation to lead, mercury and arsenic entering the food chain due to the use of phosphate fertilisers.

In developed economies, fertiliser application can be undertaken using existing irrigation systems, in a process known as fertigation. This process allows a mixture of water and fertiliser being precisely applied using a computer, tablet, or even a mobile phone, allowing applications to be controlled interactively with crop conditions. For example, this system can be set to apply water when moisture levels in the soil are below a certain threshold, and similarly fertiliser can be applied at set times after planting, particularly when young shoots are at the point they uptake nutrients. This type of system injects water and fertiliser into the soil, rather than applying it on top of the soil, making it much less likely it will be picked up by the wind, or washed off by rainwater. Unfortunately, the sophisticated digital technology required to make such a system function properly is unlikely to be available to many smallhold farmers in developing economies, at least in the short term. However, providing farmers with a simplified form of this technology along with education on plant growth cycles and appropriate fertiliser application is achievable. The key is to make the technology small enough to be able to service small farms, and affordable. The energy required to pump the water nutrient mixture to the fields could be powered by small scale alternative energy generation technology – e.g. solar, wind, micro-hydro, wave and current generated energy. The cost of the necessary digital and energy generation technology is steadily decreasing and becoming more affordable for smaller scale applications, even in more remote villages in developing countries.

The first step in realising this goal of making farming smarter is to educate governments, especially those who are subsidising the use of fertilisers, and diverting the subsidy funds to educating farming communities, which would lead to more sustainable solutions over the long run, particularly for soil nutrient management. This would reduce overall fertiliser use, without reducing crop yield, making the process more economically advantageous, as well as beneficial to human health and the natural environment.

Organic fertilisers

Organic waste, including food scraps, gardening waste, manures and even human sewage, can be treated to produce a fertiliser with both nitrogen and phosphorus, among other nutrients, that are in the correct form for plant uptake. These organic fertilisers can effectively replace mineral phosphorus fertilisers, and significantly reduce the need for synthetic nitrogen fertilisers. As previously mentioned, if not properly treated there is a risk of contamination with PTEs, so caution is suggested with the use of some sources of organic fertilisers. This approach has the added advantage of preventing non-organic fertilisers from ending up in landfills with sometimes profoundly negative impacts on the environmental. These issues are discussed in more detail in Chapter 7.

Pest control

Plants are vulnerable to attack from various forms of invasive pests, including other plants (weeds), insects that eat crops prior to harvest or during post-harvest storage, and fungi that make plants inedible. The impacts of pests on farming can be devastating. Possibly the most well-known example of this is the infamous Irish potato famine, in which one million people died of starvation, while over a million others were forced to emigrate. The population of Ireland decreased from 8.4 to 6.6 million in the course of four years (Mokyr, 2017). The pest in this case was *phytophthora* and caused almost complete loss of crops for four consecutive years. A leading cause of the problem was the fact that so many Irish farmers were growing only one or two types of potato. Pest species tend to attack particular types of crop; thus, this type of monoculture farming makes it easy for a single type of pest to quickly wipe out multiple farms.

Pest management in crop farming has been documented as far back as 2500 BC when ancient Sumerians used sulphur compounds to kill insects (Frazier, 1997). From the early 1900s the industry moved towards the use of chemical pesticides. Initially this involved the use of arsenic, lead and other poisons. This practice continued for several decades, despite a growing body of research showing that these chemicals were negatively impacting human health (Davis, 2014). The impacts were affecting both consumers, through their consumption of the produce grown in these chemicals, as well as the farm labourers who were exposed more directly during application. In the 1940s the introduction of organochlorides, such as DDT, finally saw the end of the use of known poisons. At the time these chemicals seemed like a revolution, a pesticide that seemingly didn't harm humans.

Another seeming advantage was their persistence in the environment, meaning a single application would last years. However, over time pest insects became resilient to the chemicals. It turned out that each year a small number of pests, with resilience to organochlorides would survive. These would then reproduce with other insects with the same immunity to chemical pesticides, thus eventually leading to the creation of a new generation of pests, impervious to the impacts of organochlorides. Further to this, various other negative side effects to the use of DDT were later discovered. Farmworkers, who had been exposed to DDT for many seasons, began showing medical impacts, such as cancers, and having children with birth defects. The chemicals were not targeted to the pest species, and so were also killing insects that were not pests, including some that are beneficial to plant growth, especially bees which are essential to the pollination process. DDT was also linked to health problems in other species; for example, the eggshells of some bird species became thinner, and were no longer able to produce life. This led to their numbers dwindling over time. The environmental persistence of these chemicals, once considered a bonus, is now an environmental disaster as many years after its use has ceased, significant concentrations are still being measured in the environment.

Today, most chemical pesticides take the form of organophosphates. Unlike organochlorides, these chemicals do not persist in the environment, but they are still known to have negative health impacts. Organophosphates can be absorbed through inhalation, ingestion or through the skin. Human consumers are also inadvertently ingesting organophosphates due to residue on the fruits and vegetables they eat. These chemicals have been linked to neural development problems in children, from both direct exposure and even maternal exposure during pregnancy (Mie et al., 2017).

Farmworkers are exposed to all the pathways of absorbing organophosphates as described above, especially when they fail to properly use recommended personal protective equipment (PPE). Poisoning symptoms include vomiting, dizziness and fainting. In many developing economies, especially in Southeast Asia, thousands are hospitalised each year with pesticide related illnesses (Ecobichon, 2001). Although it seems obvious that farmers should just wear PPE, many of these workers have little or no education and are not fully aware of the health risks associated with pesticide poisoning. Such protective equipment is also expensive. Schreinemachers et al. (2017) interviewed farmers from Laos, Cambodia and Vietnam, and found that while most could identify the poison symbols on pesticide labels, they had nearly all experienced symptoms of pesticide poisoning. There were also concerning beliefs, for example, that drinking alcohol after spraying would reduce the symptoms of poisoning. These sorts of misunderstandings about the dangers of pesticide poisoning, coupled with the fact that many of these regions experience warm temperatures, makes it difficult to convince farmworkers to wear heavy, uncomfortable PPE (Ecobichon, 2001).

Schreinemachers et al. (2017) also found that many farmers were not able to distinguish between insects that are pests, and those that are beneficial to plants. Even fewer were able to identify the larvae of the insects. These farmers' approach to pest control was to spray as much as possible as often as possible. Some were spraying almost daily, intending to kill off all insects, and not just pests. This also highlights the lack of awareness of pest resilience among farmers. There is also a tendency to mix different pesticides together to spray, increasing the risks to their health and increasing the evolution of pest resilience. UNESCO notes that pesticides are being used "randomly, without assessment of the pesticide formulation and quantity". (UNESCO & SCOPE, 2007)

The Rotterdam Convention on Prior Informed Consent (PIC) made it compulsory for product hazard information to be disclosed on pesticides before export or import.

The PIC gave signatory countries the right to ban any of the pesticide substances considered toxic to human health. All other signatories were required to stop exporting the banned pesticides to that country. However, 71 per cent of low income, and 55 per cent of lower middle-income countries had not banned any of the substances on the list by 2012 (Schreinemachers & Tipraqsa, 2012). Ecobichon (2001) also notes that in developing economies the use of pesticides is largely unregulated, or where it is regulated is not enforced. Pesticide use is higher in developed economies, than it is in developing economies, but looking at use over time, in developed economies it has flattened out or declined while in developing economies it is increasing at a dramatic rate (Gautam et al., 2017).

While fearing the loss of crops understandably drives many farmers to use pesticides it is possible to take an integrated pest management (IPM) approach, which could significantly reduce the use of chemical pesticides, without experiencing crop losses. IPM is not just for organic farming but in the European Union a 2009 directive requires all professional plant growers to incorporate IPM, whether organic certified or not (European Parliament and Council, 2009).

The philosophy behind IPM is that there is a threshold quantity of pests that a plant can tolerate. Below this tolerance threshold, the costs of applying pesticides outweigh the increase in crop yields thus making the use of pesticides economically unjustified. This tolerance threshold has been identified for many crop-pest combinations. IPM applies pest control responsively, i.e. after pests have been identified, rather than pre-emptively as it is in many developing economies. It also targets the specific pest, instead of aiming to kill all insects.

The advantages of IPM include:

- It's safer – farmworkers are not exposed to dangerous chemicals and there is a reduced risk of residual chemicals being present in food.
- It reduces risk of pests developing resistance to pesticides, leading to future problems with pest management.
- It is cheaper, and more effective.
- It manages pests long term, rather than just seasonally.

The components of the IPM process include (Jack, 2014):

Preparing soils

Test soils for quality, availability of nutrients, adding organic matter where necessary. Sterilising gardening tools can also reduce the risk of soil contamination leading to the development of pests.

Planting

The selective breeding of crops to create internal resilience against pests is an important area of pest management research. By planting crops that have internal resilience against pests the use of pesticides can be reduced or avoided. Selective breeding can take three broad forms (Peterson et al., 2017). The first develops a plant that can cause harm to pests, known as *antibiosis*. The second is to create a plant that directs a pest away from it, known as *antixenosis*. The problem with both of these strategies is that they to aim create both a crop response and a pest response. Similar to the use of excessive pesticides, both of these approaches could lead to the pest species evolving in a resilient way making it redundant in the future. The third option is *tolerance*, where the plant can thrive despite being attacked by the invasive species. It does not attempt to create a pest response, eliminating the possibility of the pest species developing resilience (Peterson et al., 2017). Concerns have been raised about this practice. Some types of pests can attack multiple crops, for example the corn earworm in North Carolina which may injure corn, tobacco, wild hosts, soybean and cotton (Peterson et al., 2017). If one of these crop types is bred to be resistant to the corn earworm, the pests may be diverted to the other crops in greater quantities, increasing the burden on neighbouring farms. Therefore, collaboration among farming communities on pest management is of utmost importance. Altering planting times and spacing plants differently each season can also discourage pests.

Weather

The weather can have a big impact on the outcome of crop yields. Therefore, carefully checking weather reports to ensure planting is carried out at the most appropriate time, and similarly with the application of fertilisers or pesticides, is important for maximising crop yields, while at the same time minimising damage to the environment and human health.

Monitoring

As explained above, IPM works by only applying pest control methods once pests have been found in quantities that exceed the scientifically determined tolerance threshold. Crops must be monitored regularly to check for pest quantities and damage, and only once pests are observed to be present in great enough numbers to cause economic hardship, should pest control measures be taken.

Cultural/mechanical methods

The first step should be to try using these types of methods, including removing pests by hand (for example weeding, manually picking insects off plants), or other passive techniques, like placing scarecrows in the field to deter avian and other threats to crops.

Biological methods

If cultural/mechanical methods fail to reduce pest levels below an acceptable threshold, biological control – e.g. introducing a predatory species – could be employed for controlling large and growing pest populations. This technique has been used successfully in many locations, however, it must be practised with a high degree of caution as it can have negative consequences. Even with the most up-to-date science information the unknown ecological impacts of engineering biological systems through the introduction of exotic species to ecosystems once realised can be devastating. Take, for example, the cane toad which had been used successfully in the Caribbean islands to control pest insects on sugar cane. Sugar cane was also widely grown in Australia, and in the 1930s the industry was suffering from the same pests as those found in the sugar cane fields in the Caribbean. So, the cane toad was introduced to Australia as well. Unfortunately, due to Australia's unique and delicate ecosystem, the cane toad found many other things to eat in its new home, that were easier to catch. Therefore, it did not consume the pests it was brought in to control. Furthermore, because there are no natural predators of the cane toad in Australia, its numbers have flourished and spread. Since being introduced in the sugar growing region of far north Queensland in 1935, cane toads have spread as far west as Broome, Western Australia, which is over 3000 kilometres away. Thus, the cane toad not only failed to control the sugar cane pests, it also attacked many native Australian species, reducing their numbers to endangered levels. Fortunately, many of these species have now evolved to avoid the cane toad, and some threatened species are starting to replenish. The case of introduction of cane toads to Australia provides an important lesson, the need for understanding the biology and potential threats of predatory species before introducing them into a new environment (Australian Government, 2010).

Chemical methods

Chemical methods of pest control might be employed as a last resort. For more common crops, targeted pesticides, that only attack the particular pest species, without disrupting the rest of the ecosystem, should be used where possible. Application should be minimised and follow weather patterns to ensure pesticides are used efficiently and their spread into the environment is avoided as much as possible. Chemical pesticides do not necessarily need to be harmful. There are substances permitted for use under organic farming controls in the European Union (European Parliament and Council, 2007). These fall into three categories – substances that are known to be non-toxic to humans such as spearmint oil and quartz sand; substances that are edible such as iron, potassium carbonate and rapeseed oil; and substances that are used in pest traps, and so do not come into contact with the crops (Mie et al., 2017).

Record-keeping

Perhaps the most important component of IPM is detailed record-keeping. It is important to carefully document the pests encountered and the methods used to control their quantities, so that pest control can become more systematically managed with the assistance of digital programmes. Availability of this type of information can allow farmers to identify patterns or note if and when different pest control methods are becoming less effective and be better prepared to manage pests in the future.

Many farmers are understandably concerned about trying IPM for the first time, as a loss of crop yield equates to loss of annual income, and the inability to provide for families. The loss of crop yields is not the only concern that has been raised. Farmers are also worried about the reaction of their consumers to a change in the pest management scheme (Peterson et al., 2017). Consumers tend to judge produce on visual inspection, and if the produce shows signs of pest presence, they may choose to purchase produce from a different farm. Raising awareness among consumers of how to truly assess the quality of produce, coupled with an awareness of the dangers of excessive pesticide use to their own human health and their impacts on the environment is therefore an important component of a smarter IPM programme.

Post-harvest food losses

Even after a successful harvest, the battle to eke out a living from agriculture may still not be won. Farmers still face the risk of losses during storage of crops, before they are sold or consumed. During storage produce is vulnerable to mould, especially in moist climates, to rats and other stray animals, or to insect pests that can multiply unseen in storage areas and decimate an entire harvest. The Food and Agriculture Organization of the United Nations (FAO) (Aulakh & Regmi, 2013) estimates that in sub-Saharan Africa 25 per cent of cereal crops are lost post-harvest annually, while for fruits, vegetables and root crops, this can reach as high as 50 per cent. Munang and Mgendi (2015) report that 4 billion USD worth of crops are lost during post-harvest storage each year in Africa: enough to feed 48 million people! The fear of crop losses during storage encourages farmers to sell their produce immediately following the time of harvest, when the price is lowest. In doing this, they do not retain a sufficient supply for their own use, and eventually need to purchase it back from markets when it is the time of year prices are at their highest. Thus, the farmers are losing income opportunities in multiple ways due to post-harvest storage losses.

The key to successful food storage is finding an anaerobic environment that inhibits pest breeding. Texts dating from ancient Rome document storage of grain underground, to protect it from invasive insect species, which cannot survive in low oxygen environments. In some situations, current day farmers prefer to store their produce near to living areas, so they can keep an eye on it, and ensure it is not stolen, or eaten by goats, rats etc.

The use of chemical pesticides to preserve crops during storage can be particularly detrimental to human health, as the farmers will be inhaling it, as well as ingesting it when eating the produce. Murdock and Baoua (2014) investigated various methods of pest management during post-harvest storage. They firstly considered storage in drums. While effective, the cost of the drums would be prohibitively expensive in many regions, and the produce is difficult to access when families wish to use part of it for their own consumption. The use of a triple layer of polyethylene bags was found to be the best option (see Case Study 4.2 for more details), as it is affordable, the bags cost around 2–3 USD each and can last for three seasons, and they are easy to open and reseal, making them practical storage devices for families who need to consume small amounts of grain, while waiting for the optimum time to sell the grain at markets.

Case Study 4.2 PICS bags for cow pea storage in west and central Africa (Murdock & Baoua, 2014)

Cow peas, or black-eyed peas as they are also known, are a protein rich food plant that is relatively drought tolerant and can fix nitrogen from the atmosphere to meet its own nitrogen requirements, meaning they can be grown without the use of synthetic fertilisers, even in more depleted soils. Therefore, they are an excellent crop for the low-income rural communities in arid regions, such as sub-Saharan Africa. However, they are vulnerable to bruchid beetles. These beetles multiply rapidly in favourable conditions and can consume a harvest in a very short time, quickly eliminating a family's annual income. To help farmers in Cameroon with storage, researchers from Purdue University in the US investigated four storage techniques to prevent the rapid multiplication of bruchid beetles. These methods included co-storage with ash; solar disinfestation; triple bagging in low-cost polyethylene bags; and storage in un-threshed pod form.

Another research team explored the use of drum storage in Senegal. While drum storage worked well for the Senegal region, it was not easily adaptable to other parts of Africa as the cost of the drums in other countries was prohibitively expensive (they were cheaply available in Senegal). They also made it more difficult to access the grain, as it was extracted through a small opening. Although all of the options worked well in Cameroon the researchers wanted to determine whether the storage techniques had expanded to other nearby countries. There was some evidence that they were being employed elsewhere, but this was by no means extensive. Many farmers were not using any form of storage or were using woven bags that were not airtight. Therefore, the researchers, with backing from the Bill & Melinda Gates Foundation, set about developing one of the storage techniques for further improvement. They chose the triple bagging approach as it was the most affordable and most adaptable option. The bags were sold to farmers at a cost of 2–3 USD. After overcoming initial scepticism by the farmers, the success of their use soon caught on. The concept behind the bagging system is fairly simple. It is normal for the grains to have a small portion of bruchid beetle present post-harvest. When the grain is stored in favourable conditions, the beetles then multiply and consume the grain. In the polyethylene bags, oxygen is limited and as the beetles inhale oxygen and exhale carbon dioxide, the oxygen supply continues to decrease. Without oxygen the beetles cannot reach maturity and therefore cannot reproduce. They can however lay dormant and begin to grow when oxygen is replenished. So it is important that the bags are kept air tight at all times. If the family needs to take out a portion regularly to feed themselves, they must ensure they meticulously reseal the bags every time.

Organic methods

Both soil nutrient depletion and pest invasion are exacerbated by the ever-decreasing variety of crops being grown worldwide. The simple solution would seem to simply reverse the situation and begin expanding the variety of crops being grown. Organic agricultural systems are doing just that, and thus naturally alleviating the depletion of soil nutrients and increasing pest problems through methods such as crop rotations, cover-cropping and intercropping.

Crop Rotations (Roth, 1996)

As the name suggests, crop rotation is simply alternating the crops planted each season. The science behind this method is simple: different plants require different soil nutrients, or nutrients in different forms, for their growth. Similarly, all plants will impart some nutrients to soils, helping to naturally replenish them. Therefore, alternating plants with complementary nutrient requirements can lead to natural maintenance of soil nutrition. Another way to enhance soils is to alternate between crops that have a different impact on soil structure. Planting crops with shallow roots one season, followed by deep rooted crops the following season will also maintain soil health. Rotating crops can reduce pest infestations, because different pests attack different crops, thus, pests from one season will not be able to feed the following season will start to die out, instead of continuing to thrive due to the constant supply of appropriate feed year after year.

Cover-crops

For some crops there is a gap between harvesting and planting where the soil is bare. During this time a different plant could be cultivated, which would: prevent soil erosion; improve soil structure, moisture and nutrient content; suppress weeds and other pests while attracting beneficial insects and break up compacted soils. These are known as cover-crops and their smart use can be highly beneficial, although the choice of cover-crop species is important to ensure that it is improving soil health in a way that benefits the main crop (USDA, 2018).

Intercropping

Intercropping is another simple concept – farmers grow multiple crops in the same field. From an intensification perspective, this increases efficiency by increasing yield per unit area, as complementary crops can be grown in the natural gaps of a monoculture system. Intercropping is especially good at combating weed pests. It has been argued that intercropping comes closer to emulating the way plants exist in nature. There is no natural environment where only a single species exists. To realise the benefits of intercropping the crops' species must be complementary. Similar to crop rotations, plants should complement each other in terms of nutrient use and impact on soil structure. This method has been shown to reduce disease and suppress weeds. Farmers employing intercropping techniques have also found this approach economically beneficial, due to the sales of secondary crops. Yields of the primary crop are not impacted by the secondary crop, so their sales are only extra profit.

When practised organically these three methods do not achieve optimum yields, however, their use could significantly reduce the requirement for synthetic or mineral fertilisers, and pesticides, and so their use in combination with more conventional farming methods may provide a sustainable intensification solution for farming.

Livestock management

Livestock contributes pollution to the environment in two ways. The first is through the process of microbial fermentation that occurs in the digestive system of some animals, known as enteric fermentation. Microbes break down indigestible carbohydrates and reprocess them into nutrients that can be absorbed by the animal. Methane is released as a by-product and is a very powerful greenhouse gas (GHG). The amount of methane is partially determined by the size, growth rate and production of the animal, and its use, for example as a dairy animal or one raised for meat. There are options for reducing the amount of gas and hence methane through, for example, changing animal feed type or adding supplements to the animal feed (Wei et al., 2013). Most of these options would, however, be prohibitively expensive or simply not available for many smallhold farmers in developing economies.

The second way livestock contributes pollution to the environment is their manure. Manure can be in liquid or solid form. In solid form, such as dried cow paddies, it can be often stored and later used as fuel for cooking or heating. In its liquid form it is often released into the environment untreated and can cause harm to both human health due to its content of harmful bacteria and viruses, and ecological health due to its high nitrogen concentration, which can lead to eutrophication of water bodies. Where liquid manure is stored, it is often left uncovered, leading to significant emissions of methane. However, it is possible to mitigate these emissions using a relatively simple method of covering manure lagoons with straw treated with lactic acid, which can reduce methane emissions by 25 per cent (Berg et al., 2006).

Inversely, if manure is collected and treated properly it has huge potential use as a fertiliser. Manure contains many of the nutrients required for plant growth in a form readily taken up by plants, including both nitrogen and phosphorus. Unlike synthetic and mineral fertilisers, manure helps maintain soil organic matter and soil structure. Many farmers fail to realise the fertiliser potential of manure (Teenstra et al., 2014) and its use is further discouraged in many countries, where unaware governments fail to promote its use in this way. One potential problem with the use of manure on crops is the potential for bacteria and viruses detrimental to human health to be present. These include *salmonella* and *E. coli* that cause gastrointestinal diseases. However, this is only a problem on crops that are eaten fresh, as most pathogens are killed at normal cooking temperatures. Further, treating manure through a composting process prior to its application to land will allow the biological material to degrade in a way that does not produce methane. The composting process naturally heats up to temperatures sufficient to kill pathogens and reduce the smell post-application.

Manure also has a huge energy potential. As mentioned above, solid manure is often stored and allowed to dry, then burnt as a fuel for cooking and heating. This process has many negative impacts however, as in detailed discussion in Chapter 6. A safer way to exploit the energy potential of manure is the use of anaerobic digestion (AD). This process deliberately aims to produce as much methane as possible from the manure, in a chamber that allows the gas to be captured. The gas can than provide a fuel source for many applications, including heating and cooking, electricity generation and even as a fuel for vehicles. This application will also be described in more detail in Chapter 7. After the manure has been digested in this process the residue still contains all the fertiliser nutrients. The AD process sterilises the manure, making it safe for application to land and crops. Thus, the use of AD maximises the benefit of manure by exploiting both its energy and fertiliser potential.

Disease management

Disease outbreaks among livestock can be financially devastating for farmers, as they can quickly lose their income leaving them unable to provide for their family, unable to purchase livestock for the following season, and if borrowing money to purchase livestock, they could be left in significant debt. Therefore, there is strong incentive for farmers to prevent outbreak of disease among their livestock. In conventional farming the risk of disease outbreaks is much higher than in organic animal farming due to the higher stocking density, restricted space, and barren environments in which animals are raised. Conventional animal farmers use antibiotics on all livestock, as a disease preventative measure, especially mastitis in dairy cows. The non-therapeutic use of antibiotics has been shown to improve nutrient use, energy levels and amino acid availability (Landers et al., 2012). However, use of antibiotics has also led to the emergence of antibiotic resistant "superbugs" (similar to insect pesticide resistance described previously). There is some evidence to suggest that these animal superbugs are being transferred to humans, either through direct contact with the animals and their products, or through contact with ecosystems and bacteria being transferred to water and soil (Landers et al., 2012). In organic farming, animals are given antibiotics for therapeutic management only, i.e. to treat bacterial infections, as they are in humans. The use of antibiotics to prevent the spread of disease is not necessary as organic farming allows each animal more space and they are allowed to graze. Some studies have shown a reduced risk of antibiotic resistant *salmonella* and *E.coli* in organically farmed animals (Mie et al., 2017).

There are also viral infections that can be fatal to cattle and the use of vaccinations is an option to prevent these diseases. One study in Ethiopia found that children of households that vaccinated their cattle were more likely to attend school. Improved cattle health can, therefore, have more benefits than a more reliable income (Marsh et al., 2016).

Recommendations for best practice

The goal of smart agriculture is sustainable intensification: intensifying agricultural production, without increasing damage to the environment.

In developing economies of Asia, arable agriculture has been intensified by using inefficient irrigation that wastes water, and excessive fertiliser and pesticides application leading to serious health implications for farmworkers and causing severe environmental damage. On the other hand, in Africa many farmers are not using irrigation, fertilisers or pest management strategies at all, and failing to yield enough from each harvest to support themselves and their families. This keeps them in a "poverty trap". The solution to both problems is better education for farmers. In Southeast Asia it was shown that informing farmers about better pest management practices successfully led to a reduction in pesticide use (Gautam et al., 2017). Many governments are subsidising farmers in some way, for example through heavily subsidising fertilisers that occurs in several regions of the developing world. These funds could be diverted to providing education for farmers with regard to better management practices. Many of the farms in developing countries are family businesses. Therefore, incorporating sustainable agriculture in school curricula in rural areas can potentially help to improve farming practices in these settlements. This will also prepare farmers of tomorrow for successful agricultural management that they can pass down to future generations. The pillars of this educational programme would be knowledge about managing soil moisture content, soil nutrient content and pests.

Relying on rainfall to manage soil moisture is becoming increasingly difficult as periods of drought are becoming longer and more frequent in some parts of the world. Farmers should be encouraged to employ underground or drip irrigation techniques. Especially in poorer communities, the use of perforated clay pots or even plastic drinking bottles placed strategically throughout the crop field and periodically filled to provide water during drier periods is an inexpensive (though labour intensive) way of ensuring plants have adequate water, without being wasteful. Wastewater can also be used if it is treated beforehand to minimise the risk of toxicity.

Farmers need an understanding of soil nutrient requirements for optimum plant growth. Where nutrient levels are found to be inadequate the use of manure and composted food and garden waste can be used for optimum results. The use of synthetic or mineral fertilisers should only be employed where compost is unavailable, or where the soil nutrient content is still deficient even after compost application. The timing of fertiliser application should coincide with the most appropriate point in the growth cycle and targeted to the root of the plant to avoid run-off into nearby waterways. For example, mixing fertiliser with water and applying it through a drip irrigation system is ideal. Where timed application of fertilisers is not feasible, the use of slow release fertilisers, or inhibitors can be used.

Farmers should be supported in the use of an integrated pest management system. Where chemical pesticide use is seen to be still necessary, farmers should be taught about the health risks associated with pesticides and encouraged to properly use personal protective equipment. Teaching farmers to keep detailed records of their activities will help them to recognise patterns in pest emergence and plant response to fertiliser use. This can help them build up a better understanding of when to apply chemicals and how much to use.

Farmers should also be encouraged to employ techniques of crop rotations, cover-crops and intercropping where possible to naturally support soil nutrient management, discourage pest invasions and produce more diversity of crops and often for more money.

Livestock management

The key to sustainable livestock management is integrated manure management. The first step to achieving this is to encourage the use of manure for fertilisation. Where possible this can be linked to energy production derived from manure through anaerobic digestion, with subsequent application to land to improve soil nutrient content. Animals should be given adequate space to naturally support their growth and reduce the risk of disease and the use of vaccinations should be encouraged. However, the use of antibiotics should be responsive rather than preventative to prevent the evolution of disease resilient superbugs.

References

ADAS 2009. Phosphorus: A vital non-renewable resource. Durham: Premier Waste Management.

Aulakh, J. & Regmi, A. 2013. Post-harvest food losses estimation: Development of consistent methodology. FAO.

Australian Government 2010. The cane toad fact sheet. Canberra: Department of the Environment, Water, Heritage and the Arts.

Berg, W., Brunsch, R. & Pazsiczki, I. 2006. Greenhouse gas emissions from covered slurry compared with uncovered during storage. *Agriculture Ecosystems and Environment*, 112, 129–134.

Davis, F. R. 2014. *Banned: A history of pesticides and toxicology*. New Haven: Yale University Press.

Dwivedi, S. L., Bueren, E. T. L. V., Ceccarelli, S., Grando, S., Upadhyaya, H. D. & Ortiz, R. 2017. Diversifying food systems in the pursuit of sustainable food production and healthy diets. *Trends in Plant Science*, 22, 842–856.

Ecobichon, D. J. 2001. Pesticide use in developing countries. *Toxicology*, 160, 27–33.

European Parliament and Council 2007. Council regulation 834/2007/EC on organic production and labelling of organic products. *Official Journal*, L189, 1–23.

European Parliament and Council 2009. Directive 2009/128/EC establishing a framework for community action to achieve the sustainable use of pesticides. *Official Journal*, L309, 71–86.

FAO 2002. *Crops and drops* [Online]. Rome: FAO. Available: www.fao.org/docrep/005/Y3918E/y3918e10.htm [Accessed 15 March 2018].

FAO 2017. *FAOSTAT* [Online]. Rome, Italy: Food and Agriculture Organization of the United Nations. Available: www.fao.org/faostat/en/#home [Accessed 19 July 2018].

France 24 2017. *Dried-up Aral Sea springs back to life* [Online]. YouTube. Available: www.youtube.com/watch?v=pk4v0uu5rkY [Accessed 10 September 2018].

Frazier, M. 1997. *A short history of pest management* [Online]. PennState Extension: PennState. Available: https://extension.psu.edu/a-short-history-of-pest-management [Accessed 15 February 2018].

Gardiner, D. T. & Miller, R. W. 2008. *Soils in our environment*. New Jersey: Pearson Prentice Hall.

Gautam, S., Schreinemachers, P., Uddin, M. N. & Srinivasan, R. 2017. Impact of training vegetable farmers in Bangladesh in integrated pest management (IPM). *Crop Protection*, 102, 161–169.

Hertel, T. W. & Baldos, U. L. C. 2016. Attaining food and environmental security in an era of globalization. *Global Environmental Change*, 41, 195–205.

Hunger Notes 2016. *2016 world hunger and poverty facts and statistics* [Online]. Washington DC: Hunger Notes: World Hunger Education Service. Available: www.worldhunger.org/2015-world-hunger-and-poverty-facts-and-statistics/ [Accessed 15 February 2018].

Jack, C. 2014. Integrated Pest Management. University of Liverpool.

Khoury, C. K., Bjorkman, A. D., Dempewolf, H., Ramirez-Villegas, J., Guarino, L., Jarvis, A., Rieseberg, L. H. & Struik, P. C. 2014. Increasing homogeneity in global food supplies and the implications for food security. *Proceedings of the National Academy of Sciences*, 111, 4001–4006.

Landers, T. F., Cohen, B., Wittum, T. E. & Larson, E. L. 2012. A review of antibiotic use in food animals: Perspective, policy, and potential *Public Health Reports*, 127, 4–22.

Makombe, G., Namara, R. E., Awulachew, S. B., Hagos, F., Ayana, M. & Kanjere, M. 2017. An analysis of the productivity of technical efficiency of smallholder irrigation in Ethiopia. *Water SA*, 43, 48–57.

Marsh, T. L., Yoder, J., Deboch, T., McElwain, T. F. & Palmer, G. H. 2016. Livestock vaccinations translate into increased human capital and school attendance by girls. *Science Advances*, 2.

Mathur, V. B., Verma, A., Dudley, N., Stolton, S., Hockings, M. & James, R. 2006. Opportunities and challenges for Kaziranga National Park, Assam over the next fifty years. UNESCO.

Mie, A., Andersen, H. R., Gunnarsson, S., Kahl, J., Kesse-Guyot, E., Rembiałkowska, E., Quaglio, G. & Grandjean, P. 2017. Human health implications of organic food and organic agriculture: A comprehensive review. *Environmental Health*, 16, 22.

Mokyr, J. 2017. Great famine. *Encyclopaedia Britannica*.

Munang, R. & Mgendi, R. 2015. Africa's broken food security value chain: unravelling the hidden fortune. *This is Africa*. 20 August. Available: https://thisisafrica.me/africas-broken-food-security-value-chain-unravelling-hidden-fortune/ [Accessed 11 September 2018].
Murdock, L. L. & Baoua, I. B. 2014. On Purdue improved cowpea storage (PICS) technology: Background, mode of action, future prospects. *Journal of Stored Products Research*, 58, 3–11.
Murtaza, G., Usman, Y., Niazi, N. K., Usman, M. & Hussain, T. 2017. Bioaccumulation of potentially toxic elements in cereal and legume crops: A review. *CLEAN – Soil, Air, Water*, 45, 15.
Peterson, R. K. D., Varella, A. C. & Higley, L. G. 2017. Tolerance: the forgotten child of plant resistance. *PeerJ*.
Roth, G. W. 1996. Crop rotations and conservation tillage. University Park, Pennsylvania: The Pennsylvania State University.
Rurak, M. 2012. Advantages and disadvantages of controlled-release fertilisers. Madison, Wisconsin: Department of Soil Science, University of Wisconsin, Madison.
Schreinemachers, P., Chen, H.-P., Nguyen, T. T. L., Buntong, B., Bouapao, L., Gautam, S., Le, N. T., Pinn, T., Vilaysone, P. & Srininvasan, R. 2017. Too much to handle? Pesticide dependence of smallholder vegetable farmers in Southeast Asia. *Science of the Total Environment*, 593–594, 470–477.
Schreinemachers, P. & Tipraqsa, P. 2012. Agricultural pesticides and land use intensification in high, middle and low income countries. *Food Policy*, 37, 616–626.
Shafiee, S. & Topal, E. 2009. When will fossil fuel reserves be diminished? *Energy Policy*, 37, 181–189.
Stamoulis, K. & Lipper, L. 2012. Agricultural input subsidies and the green economy: Fertilizer subsidies in Sub-Saharan Africa. Rome, Italy: Food and Agriculture Organisation of the United Nations.
Struick, P. C., Kuyper, T. W., Brussaard, L. & Leeuwis, C. 2014. Deconstructing and unpacking scientific controversies in intensification and sustainability: Why the tensions in concepts and values? *Current Opinion in Environmental Sustainability*, 8, 80–88.
Teenstra, E., Vellinga, T., Aektasaeng, N., Amatayakul, W., Ndambi, A., Pelster, D., Germer, L., Jenet, A., Opio, C. & Andeweg, K. 2014. *Global Assessment of Manure Management Policies and Practices*. Wageningen: Wageningen UR Livestock Research.
UNESCO & SCOPE 2007. Human alteration of the nitrogen cycle: Threats, benefits and opportunities. Paris, France: UNESCO-SCOPE.
United Nations 2015. *Sustainable development goals* [Online]. United Nations. Available: www.un.org/sustainabledevelopment/ [Accessed 1 February 2018].
USDA 2018. *Cover crops* [Online]. Available: https://plants.usda.gov/about_cover_crops.html [Accessed 15 February 2018].
Wei, M., Greenblatt, J. B., Donovan, S. M., Nelson, J. H., Mileva, A., Johnston, J. & Kammen, D. M. 2013. California's Carbon Challenge Phase II Volume I: Non-electricity sectors and overall scenario results. *Scenarios for meeting California's 2050 climate goals*. Berkeley, California: California Energy Commission.
White, R. E. 1997. *Principles and practise of soil science*. Carlton: Blackwell Science Ltd.

5 Housing

Introduction

This chapter explores the concept of smart housing in rural areas. As many rural communities modernise, and are exposed to the effects of globalisation, the lifestyles of their inhabitants are likewise changing (Bański & Wesołowska, 2010, Gkartzios & Scott, 2010, Kirk, 2001). Such changes are challenges most evident in the construction of houses in rural areas. The concern here is with housing in rural areas in developing countries, which differ from housing in rural areas in developed countries in a number of ways across socio-economic, cultural, religious and environmental spheres. In essence, everyday life in rural areas differs with the context that is tied to the land and its geography. A rural setting often implies a core settlement situated amidst, or adjacent to, peripheral agricultural lands (Fuentes, 2010, Nepal, 2007). Geography and the distances between rural settlements significantly influence the spatial arrangement of different elements within villages (Nguyen et al., 2011, Lim, 2007, Anna-Maria, 2009). In general, the economic structure of rural areas is mainly dependent on primary industries, such as agriculture, that rely on local natural resources. The relative isolation of villages also means that there are limitations on the construction technologies and building materials that are locally available and can be sourced. In such settings it is imperative to examine how "smart" houses would look and function.

The interest here is in examining the salient features of rural houses, which are often described as "vernacular" or "traditional" forms of architecture, which have evolved in response to local environmental and social conditions (Zhai & Previtali, 2010, Bodach et al., 2014, Gautam, 2008). How can housing in rural settings best adapt in the face of urbanising forces that are rapidly changing the landscapes in which many rural settlements are located? When we speak of housing we mean more than just houses as single entities but the total environment and habitation system built around networks of activities in which rural inhabitants are engaged (Rapoport, 1998). Therefore, an exploration of rural housing by looking at socio-economic and cultural dimensions is necessary when the final goal is to achieve sustainable rural development (Chiu, 2004).

In seeking to explain what smart housing would mean in the context of a Smart Village, this chapter first examines three prominent threads around rural housing

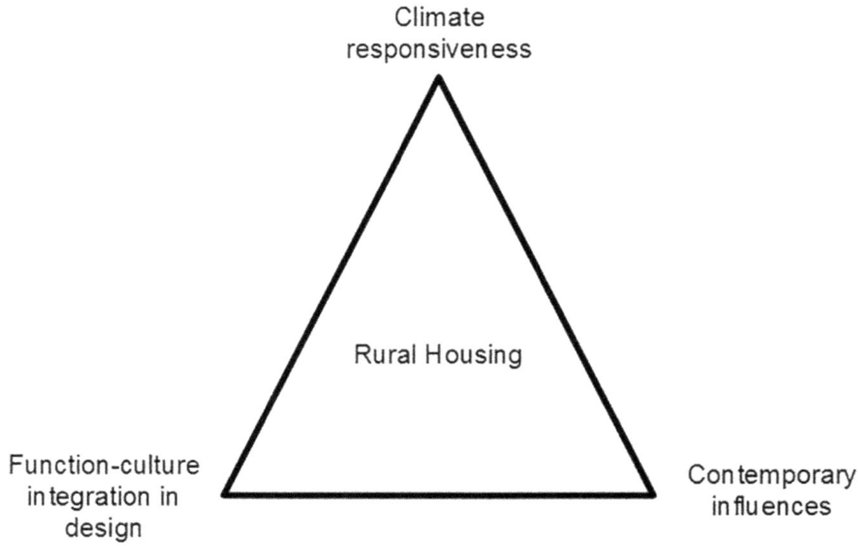

Figure 5.1 Key threads around rural housing

as illustrated in Figure 5.1: climatic responsiveness as a distinguishing feature of a rural house; function-culture integration in design of rural houses and contemporary influences on rural houses and transformations of rural landscape. As new materials and technologies make inroads into rural areas, concepts of shelter and house are also changing. The new rural landscape has its own demands for better living conditions and these are often to their detriment by relinquishing their vernacular identities. More importantly, how contemporary smart approaches to housing intersect with the vernacular and traditional is very important and the central theme of this chapter.

Rural house and climatic responsiveness

Houses in rural areas are generally referred to under the broad rubric of "vernacular" or "traditional houses". Vernacular buildings are generally built using indigenous building traditions (Vellinga, 2006), involving local builders and residents, employing locally available materials, and in a way that responds to climatic conditions (Foruzanmehr & Vellinga, 2011, Abu-Ghazzeh, 1997, Dodo et al., 2014, Kimura, 1994, Maudlin, 2010). For residents, these structures "in every way reflect their belief systems" (Cromley, 2008) and cater to the various economic, cultural and social needs (Singh et al., 2011). The vastness of the study of vernacular architecture can be seen in the way Dell Upton (1983) has conceptualised the field and suggested four "avenues of inquiry": 1) *Object-Oriented Studies* concerned with the buildings themselves and how they are made and change over time; 2) *Socially Oriented Studies* where "buildings are being examined both as a part of everyday existence" using a variety of evidences; 3) *Culturally Oriented Studies* which examine cognitive structures of a social group in producing specific kinds of architecture; and 4) *Symbolically Oriented Studies* that deal with the symbolic character of the architecture. This chapter is not a place for a review of the voluminous literature that exists about vernacular architecture. Rather, our interest is in examining characteristics that are associated with vernacular houses that have direct bearing for contemporary smart approaches.

In rural areas, the agricultural character of the territory often leads to "a varied catalogue of traditional rural buildings" that are unique to those social groups inhabiting the land (Fuentes, 2010). In case of European rural landscapes, Fuentes (2010) identifies some as "subterranean wine cellars, traditional wineries, dovecotes, sheep houses, water mills or traditional beekeeping structures, among others". In case of Sukur in Nigeria, the stone structures in the form of houses, farm terraces and walkways are the most distinct feature of landscape (Dodo et al., 2014). Fuentes (2010) notes that "the specific techniques, materials and building details used in each particular rural area turn its vernacular architecture into an identity symbol and rich ethnographic evidence about rural technologies and ways of life in the countryside." Although rural houses take on a diversity of forms, their primary function continues to be of some use in the agrarian landscape and to work with the climate they are in.

At the core of constructing a rural house is the need to seek shelter from the climatic situation using available resources that are often limited. Correspondingly, traditional building practices have relied heavily on sound knowledge of climate so as to harness nature's energy – mainly from sun and wind to create comfortable living conditions. The material and technology to build is also rooted in the local context – accessibility, availability and functionality. Since these are generally a part of the same geography their suitability in achieving comfort without any mechanical aid is higher. It is through these practices that rural houses have been able to deal with energy issues. While the final form and architectural expressions of shelter depends on geographical diversity, social, cultural, religious beliefs and values of residents, the main characteristic of a rural house continues to be its climatic responsiveness.

Evaluating climatic responsiveness of vernacular houses seems to have caught up with the research around built environment and energy. In recent years, several studies have taken up environmental performance of buildings (Dili et al., 2010, Nguyen et al., 2011, Indraganti, 2010, Singh et al., 2009, Shastry et al., 2014): in fact, an entire journal is dedicated to such studies of traditional architecture. Here, we provide an overview of climatic analysis of a typical rural house.

Of the many features assessed for climatic performance of a rural house, the following are more frequently used: "settlement pattern, building form and orientation, building stories and internal space arrangement, design and construction materials of walls, roof, foundation, floors, ceilings and openings" (Bodach et al., 2014).

Bodach et al. (2014) conducted an expansive study of different typologies of houses and settlements across four climatic zones of Nepal: Sub-tropical; Warm-temperate, Cool-temperate and Alpine; and Tropical. They analysed the above-mentioned eight characteristics of houses in these zones and found that the traditional architecture in all settlements was "very well adapted to the local climate conditions" as the main strategies included "enhancing air movement within the building and protecting from the strong solar radiation" (Bodach et al., 2014). In case of rural houses in Vietnam, Nguyen et al. (2011) found that such passive design strategies and solutions are simple, inexpensive and easy to apply, and demonstrate "a deep understanding of the building and its surrounding environment".

That radiation and ventilation are the most important considerations for achieving comfortable living conditions in rural areas is also evident from another detailed study of vernacular houses in other climatic zones. This is illustrated by Gautam (2008) in the case of the central Indian state of Jharkhand, where two different styles of houses were examined. Gautam builds on to the definition of vernacular architecture forwarded by Rapoport (1990) on the basis of process characteristics and product characteristics and makes distinctions between two types: huts and havelis (large mansions). He observes that the huts were "self-built by family members" with materials such as mud, sticks, grass and pebbles and "their modest beauty lies in being less influenced by self-conscious decorative attempts than from pure, practical shapes produced by adapting local material as economically as possible to mitigate hostile environmental elements" (Gautam, 2008). In comparison, the architecture of havelis evolved from the desire of the merchant class to display their wealth: these were built by local craftsmen using burnt clay brick and were highly influenced by western motifs. Gautam argues that "the buildings that resulted from this process are quite different from the earlier type, combining traditional with modern, but they are still vernacular because they are the outcome of local vernacularization of modernity" (2008). The most significant feature, however is that both these typologies of houses were "constructed, without any mechanical means, in such a manner as to create micro-climates inside them to provide high thermal comfort levels" which is the necessary climatic condition in the region (Gautam, 2008).

Regional variation of climates is often reflected into a range of climate-responsiveness features that become the defining characteristics of vernacular houses. This is demonstrated in many studies from the western coastal state of Kerala in India. In a detailed assessment, Dili et al. (2010) bring out the importance of usage of courtyards and how placing rooms around it help to provide thermal comfort to the occupants. In their investigation they focus on how the indoor environment is created through orientation of the building, internal arrangement of spaces, the presence of a courtyard, use of locally available materials and special methods of construction. Through their analysis they show how the house forms "offer flexibility for the inhabitants to shift their activities from one place to other in different climatic conditions" (Dili et al., 2010).

Different climatic zones have different parameters that become important in design considerations. While cross-ventilation appears to be the most important objective in coastal regions, dealing with heat takes precedence in considerations for rural houses in hot and arid climatic zone. In a study of the vernacular settlement of Marikal in the dry plateau of Andhra Pradesh, Indraganti (2010) examines the climate appropriateness of various features of a house and settlement. It focuses on environmental elements such as heat, humidity, air movement and light and the general activity pattern of the residents and examines passive design features grouped under eight headings: layout, spacing, air movement, openings, walls, roofs, outdoor sleeping and rain protection (Indraganti, 2010). In Marikal, the primordial consideration in development of form and site planning is protection from dry summer weather, which evolved "as a tightly knit fabric of small housing clusters" (Indraganti, 2010). The analysis suggests how using traditional building practices ties bioclimatic concepts with the living styles, landscape and relies on "little wastage of resources" (Indraganti, 2010).

In sum, it can be said that houses and housing patterns in rural areas are generally responsive to climatic conditions as through passive design principles they achieve environmental comfort. However, it should also be noted that the direct cause-effect relation in incorporating climatic features in rural houses may be overrated. Nguyen et al. (2011) in their detailed study of Vietnamese rural houses have indicated that "not all vernacular buildings have perfect building physics". Some scholars have argued that it is more of cultural dimensions including social hierarchies, customs and traditions that led to construction of vernacular dwelling without an explicit focus on climate-responsiveness (Yıldırım & Turan, 2012, Singh et al., 2009). Additionally, there may be other imperatives such as social hierarchies or economic conditions rather than climate determinism that is often associated with local culture of housing (Cromley, 2008). Often, knowingly or unknowingly climate-oriented design features become a part of cultural traditions that become central to house design and vice versa.

Building designs

Layout, design features, character and fit with landscape

Beyond climatic responsiveness, the constituting elements of architecture in rural houses have multiple functions and meanings. The traditional design principles in rural areas have evolved around the need to support agricultural productivity of households and rural lifestyles that demand flexible and functional spaces. Lim (2007) puts it succinctly: "vernacular architecture accommodates cultural needs".

Owing to agriculture related livelihood activities, rural residents emphasise the need for outdoor spaces such as courtyards and semi-covered spaces such as verandas in their houses. Dili et al. (2010) show how these are integral to vernacular houses in Kerala where they are used for a variety of both work and leisure activities.

Often vernacular houses reflect "an agrarian way of life and domestic space" (Lopez, 2010) and therefore one finds courtyards are the basic unit of house design across rural societies. In "Assam type" houses in the north-eastern parts of

India, the courtyard provides the necessary space for post-harvest activities like drying, winnowing and weaving and also satisfies the social and cultural needs (Singh et al., 2009). Lopez (2010) noted the heightened functionality and meaning acquired by courtyards through a study of houses in San Miguel, Mexico.

Besides the functionality, a courtyard also has an important climatic purpose. The courtyards in Assam type houses act as transition spaces that can be "modified according to the needs in summer and winter months to maintain comfortable conditions inside the building" (Singh et al., 2009). Nguyen et al. (2011) found similar climatic role for courtyards in Vietnamese vernacular houses where "building courtyard played a significant role on ventilation flow rate of the rooms facing the courtyard". In Iran, courtyards provide passive cooling and were indispensable elements of the vernacular architecture (Foruzanmehr & Vellinga, 2011).

On the outside, the presence of a courtyard contributes to some uniformity in the streetscape due to shared facade elements which also mean a "marked continuity and homogeneity of appearance" (Lopez, 2010). The cluster planning around courtyards is of high importance not only for climatic reasons but also for cultural connections it helps to foster within the community. In the study of Marikal, the courtyard is used for a variety of purposes depending on the household's occupation. Use and maintenance of the courtyard is shared by all households, and the space can also act as a wedding venue (Indraganti, 2010).

While it is generally agreed that passive design principles lead to emergence of a housing form that broadly responds to rural environments, it is also observed that traditional houses pose some challenges for their inhabitants. These are particularly related to health risks and hazards associated with energy use in rural areas. There continues to be a high dependency on traditional biomass fuels like wood, dung and crop residues amongst villagers to meet their energy needs (Joon et al., 2009). It is in this regard that traditional designs are considered poor as they pose challenges related to respiratory diseases and other ailments.

In rural areas, a house is also a cultural proposition not only for an individual but also for entire villages. In the words of Lim (2007):

> Households are bound together in a village by the sharing of services, religious rites and obligations. The construction of a house is a task shared by the community. The sense of identity of the house owner is materialised in the construction of her house as endorsed by the entire community.

Besides performing the basic functionality, a rural house represents a bundle of meanings related to culture. "Culture" is a theoretical construct and "Cultural" refers to *ideational* variables but these are manifested through social expressions "as family and kinship structures, social networks, roles, statuses, social institutions, and the like" (Rapoport, 1998). Following Rapoport's explanation of importance of culture in housing design, it could be argued that the social variables – the activity systems that houses seek to support – and lifestyle, help in executing choices and forms of housing in rural areas. Understanding culture as a way of life it seems prudent to accept that rural areas do change, including rural housing.

Changes over time

Local materials and micro-enterprise; cultural adaptation of tradition

As the world urbanises at a pace faster than witnessed ever before, dramatic changes are also being observed in rural areas. Figure 5.2, for example, shows how a traditional house in Assam, India is being influenced by a contemporary RCC framed building system. There is considerable literature around how modernity is making its way even into the remotest areas (Kirk, 2001, Bański & Wesołowska, 2010, Kumar et al., 2010). Rapoport (1998) remarks how "developing countries in particular provide clear and striking examples of rapid changes in values, life-styles, activity systems, kinship and family structures, roles, institutions, social mechanisms, social networks, meanings, and images (and hence wants)".

For this chapter, we focus on the changes most relevant to rural housing. These changes can be categorised into the following clusters of observations: 1) contemporary design approach to houses; 2) advent of new building materials and construction technology; 3) modifications to sociocultural and religious meanings; and 4) changing dynamics of energy use.

1) Contemporary design approach to houses

The changes in approaches to the design of rural houses are influenced by many sources. These include urban imagery brought by returning migrants, access of rural households to global television networks, information channels, etc., rural-urban market integrations, and in general rural development initiatives that have also led to changing demographics and gender relations.

One of the main reasons for changing rural fabric is rural residential mobility (Gkartzios & Scott, 2010). The processes of rural in-migration and counter-urbanisation appear to have strongest influences on the emerging forms of housing. Gkartzios & Scott (2010) provide a detailed explanation of type and

Figure 5.2 Vernacular housing on Majuli Island, Assam, India

motivations for new house building in rural areas using two strands of housing forms: *urban-generated rural housing* where housing sought in rural areas by people living and working in urban areas and thus the development is driven by urban centres; and *rural-generated housing that reflects* "housing needed in rural areas in the established rural community by people working in rural areas or in nearby urban areas who are an intrinsic part of the rural community by way of background or employment". In their case studies from rural England, they found social and physical characteristics of rural living (as against the peril of urban areas) as a very important reason for people to move to a rural locality.

In rural communities, there is also increasing evidence of house-building activities by returning migrants. This is illustrated by two studies from Mexico. In one study, Lopez (2010) explains how migrants not only "build for retirement" but also in the process "define themselves as successful to their family and community, or as a means of self-expression". In another study, Pauli (2008) focuses on houses in rural areas that were largely conceived and executed by women who stayed back in the villages while their spouses migrated to cities for work. Houses are classified into three types using the following classification criteria: building material, number and use of rooms, and inner and outer decor. The three types were: houses made of local materials [adobe, teja and roofing tiles]; houses using a mix of materials [corrugated iron sheets and cement available from nearby markets]; and "an urban, popular design" where the roofs and walls are made of cement, outer walls painted in bright colours and houses have an average of 3.5 rooms with each room having a different function. The third category was on the rise and the "living room" was becoming a central feature of the house. She notes how creative women involved in designing these houses "got much of their inspiration from television, magazines, and work as maids in Mexican cities" (Pauli, 2008). A similar approach to building new vernacular houses with modern designs and newly available materials was observed in Indonesian villages (Vellinga, 2006).

With an urban imagery making inroads in rural areas, the fabric of the place and spatial relations are bound to be altered. The desire to build modern houses with modern floor plans and multiple stories meant zealous migrants broke down old houses and in doing so they ruptured the continuous streetscapes (walls and rooflines) that were once a characteristic of the village (Lopez, 2010). Similar to the exterior, the interior of houses has changed significantly as the idea of "courtyard" which was central to a rural house is displaced by seemingly space-saving and enclosed functional spaces. The abandoning of the courtyard causes disruption to several of the social, cultural and ritual activities that were once integral to the rural lifestyle. But it is possible that this is what residents want in terms of development and may be what they identify with (Vellinga, 2006, Shastry et al., 2014).

The changes in approaches to house design, while impacting individual households also has significant influence on the community. Vellinga (2006), based on Indonesian case studies, notes that "in the past construction of vernacular housing involved whole families working together under a master builder" and "incorporated rituals and social festivities aimed at enhancing the house's vitality and fortune."

Rural mobilisation has a strong impact on the disjointedness and rupturing of the fabric of villages. While the building of new, modernised vernacular housing (like its traditional predecessors) serves "as an objectification of the wealth of its owners" (Vellinga, 2006), it has also become a precedent and guide for residents. The symbolic importance of such new houses cannot be undermined. This has been explicitly noted by Indraganti (2010) in her study of Mirakal where "villagers attach a tag of higher social status and modernity to a concrete roof house than to a tiled vernacular house".

Aspirations for housing change with an overall and significant socio-economic rise in rural areas and so does the use and meaning of houses. Shastry et al. (2014) make a valid point: "owing to the recent availability of electricity, transport facilities, improving education facilities in villages and job opportunities in the cities, the urban influence on the current generation is increasingly evident in the lifestyles of people".

2) Modifications to sociocultural and religious meanings

Several studies draw attention to how building a house is not only a construction activity but also constitutes a part of the social, cultural and religious context of rural areas (Rapoport, 1987, Cromley, 2008, Lim, 2007, Rapoport, 1998). The design of rural houses is strongly embedded in cultural practices. In most traditional design practices many elements of a house and settlement are enshrined in religious beliefs and value systems. For instance, the orientation of a house is often based on the ancient science of *vaastu shastra* – particularly the direction of main areas and serving areas is governed by such rules; scale and proportions are defined which lead to clear distinctions between private areas, spaces of public interactions, gendered use of space, and so on.

As discussed above, the substitution of the courtyard has very strong impacts on cultural spaces and cultural meanings in everyday lives. Similarly, when traditional spaces in rural houses are modified with the modern trappings of an urban home a new language emerges. To understand new rural houses, a new vocabulary and syntax becomes necessary. This is often at a crossroads with traditional meanings.

3) Advent of new building materials and construction technology

The aspirations of villagers for building new houses are supported by changing market dynamics. In rural markets, access to and availability of industrial construction materials such as concrete and steel has improved considerably (Shastry et al., 2014) altering the ways rural houses are built. The desire to build new houses that mimic urban lifestyles also influences choices to go in for faster construction and saving of space. Shastry et al. (2014) noted some of these changes while examining the environmental performance of old versus new building in the village of Suhangatti including the stone chips and mortar that are freely available in the local area being replaced by burnt clay bricks and cement from nearby towns; and the heavy (>450mm thick) rubble masonry walls have slimmed down to slender (<230mm thick) brick masonry which has lower heat storing capacity.

Material substitution in construction has implications for climatic performance and consequently the internal comfort conditions. The replacement of the thicker walls made of natural materials by thinner material such as lightweight bricks reduce the thermal mass of houses leading to less comfortable living conditions. In Suhangatti, (Shastry et al., 2014) simulations of modern materials on vernacular houses found that "adopting modern building material and design configurations would increase indoor temperatures by 7° to 10° C in summer leading to thermal discomfort". This also forces people to rely on mechanical means for ventilation and incorporate those features in the house design (Indraganti, 2010).

The advent of new materials and technologies has a crucial impact on the traditional wisdom of construction and local construction skills and crafts. In many cases, it is observed that the traditional modes of construction lose patronage in favour of new materials and technologies and subsequently wither. Citing examples from a case study of vernacular houses in a south Indian village, Indraganti (2010) notes how the local roofing craft of tiling is almost extinct and the local potter community has shifted to making other commercially viable products, while the local stonecutter community, who used to work on stone walls and roofs has moved elsewhere. However, new constructions may eventually lead to innovation and revitalisation. This is most evident in the case of Mexican villages where locally crafted goods that complement the lavish aesthetic of the new vernacular lead to an expansion of the market for local crafts "locally made custom doors and windows and wrought-iron ornamentation are still preferred to prefabricated windows and doors" (Lopez, 2010). However, as demand increases for locally produced crafts they become more expensive. Similar respect for traditional crafts has been noted by Vellinga (2006) in Indonesian examples.

One area where new materials and technologies have begun to show significant impacts is energy use in rural housing. Conventional housing construction in rural areas had low energy consumption in the "procurement, manufacture, processing and recycling of building materials; in transporting building materials to the building site; and in on-site construction activities" (Tiwari, 2001). Moreover, the cost of shelter was determined mainly by the availability of building materials which had a built-in mechanism of efficient use of resources at low cost. However, this is all set to change in contemporary rural housing as more and more material is readily available at a reasonable cost because of economies of scale and distribution in rural areas. Tiwari (2001) warns the rapid dissemination of mass produced homes and buildings designed by external teams of architects and engineers is threatening the vernacular traditions that help define the culture of a people and a place. While the threat to vernacular traditions is always imminent as new materials and technologies evolve in housing construction, there are also real opportunities to work with newly available materials alongside traditional passive design principles for achieving better comfort and environmental sustainability.

4) Changing dynamics of energy use

The relationship between rural housing and microclimate in rural areas has another significant dimension – that of building construction as it involves energy consumption for acquiring raw building materials and generation of waste during construction which indirectly emits greenhouse gases (Kumar et al., 2010). Case Study 5.1 describes an approach used in Australia to improve the environmental performance of buildings.

Case Study 5.1 Sustainability certification programmes

The importance placed on the sustainability of land development, of all types and scales, in many countries, is an acknowledgement that land development consumes an enormous amount of the earth's resources in the form of materials, energy and water, and negatively impacts the natural environment and its ecology at development sites. Buildings themselves, both their construction and operation, are responsible for a significant amount of greenhouse gas (GHG) emissions. Buildings consume close to a third of the world's energy (Lucon et al., 2014). Energy used for cooking, cleaning, lighting, heating and other household activities in homes in the United States, for example, accounts for some 20 per cent of the country's total energy consumption (Kaza, 2010).

The Urban Development Institute of Australia's Envirodevelopment (ED) programme has developed an assessment procedure for assessing the sustainability credentials of land developments. This procedure is composed of six ED assessment "leaves", which pertain to the sustainable use of water and energy, handling of waste, selection of materials, conservation and enhancement of ecology and the building of communities. Developers seeking ED certification must demonstrate how their project meets the assessment criteria for at least four out of six of these "leaves", however, many developers try to acquire all six leaves. These six ED leaves include:

Ecology

The aim of this leaf is protection and enhancement of native ecosystems and minimising degradation caused by the development and to encourage plant and animal species to live in harmonious and balanced environments. The focus here is on conservation and/or establishment of healthy, biodiverse ecosystems.

Materials

This leaf is concerned with recycling and minimising pollution from construction materials. To acquire this leaf materials used in construction have

(continued)

(continued)

to be recycled, reused, renewable, non-toxic, locally sourced or a combination of these initiatives and have low life cycle energy requirements.

Energy

The aim of this leaf is to ensure implementation of measures to achieve energy use reduction and associated reduction in GHG emissions across a development. To achieve this leaf a development must show that it reduces GHG emissions through energy efficient appliances and fixtures, passive cooling and heating designs and importantly the use of alternative energy sources (e.g. solar, wind, biomass gas, hydro).

Water

This leaf concerns improving water use efficiency, conserving potable water and encouraging the use of alternative and recycled water sources. This includes such measures as using low water-needing plants for crops and landscapes, water saving fixtures and fittings within dwellings and use of recycled water wherever possible.

Waste

This leaf concerns the recycling of waste, including of construction and demolition materials, and minimising pollution. The aim is to reduce the amount of waste going to landfills. This also includes promoting awareness in residents of the need to reduce, and efficiently use resources, as well as to encourage recycling, composting, along with providing appropriate recycling facilities.

Community

A sense of community is a key component to the success of any rural development project. A sense of community means an active and healthy lifestyle for the residents and adequate local facilities and services to support such lifestyles. Achieving this leaf involves showing that a development has undertaken community consultation and provided local facilities and employment that are both safe and accessible. This leaf is also concerned with protecting the heritage of places and maintaining and enhancing community assets.

These six sustainability assessment leaves can be very useful in guiding the design and planning of new rural development with the aim of making them optimally environmentally sustainable.

In the last few decades, electrification in rural areas has significantly contributed to overall socio-economic development and changing dynamics of energy use and energy security in rural housing sector. In Marikal, Indraganti (2010) found that increasing availability of ceiling fans saw many previously outdoor activities such as cleaning grain, resting and reading move indoors. Subsequently, windows are becoming bigger to naturally increase indoor lighting. (The region has limited electricity access so indoor lighting is not constantly available.)

The availability of energy options is also leading change. For instance, cooking energy is the major share in total household energy consumption and availability of cooking fuels at affordable prices remains a major concern for the large population living outside the formal energy system (Laxmi et al., 2003). Access or lack of it to energy sources generates hardships, negative externalities of pollution and gendered division of labour. Therefore, integration of energy use is something that needs careful consideration in the design of rural houses. The changes related to energy use and houses are going to be even more dramatic as smart technologies make their way on to the rural scene. Joon et al. (2009) rightly point out that the picture of energy use is "often complex as in many rural areas, households often employ a 'multiple model' of stove and energy use in which households stretch across two or more steps of the energy ladder and fuel substitution is often partial". In sum, Vellinga (2006) position aptly applies to rural houses and housing where building traditions are in "an ever changing continuous interplay of precedent and innovation that dialectically responds to changes in society, culture, and the natural environment". This interplay now has a new player – the idea of "Smart Villages".

Smart Villages

The "smart" in current discourses of development is closely related to access to energy alongside the integration of modern information and communication technologies in improving living conditions. Extending this to rural areas means that a "Smart Village" appears to be "a rural analogue of the 'smart city' concept" (Holmes et al., 2015). Many governments see Smart Village as "an innovative means of improving rural people's life" (Abdul Razak et al., 2013) where technology acts as a catalyst for development (Sai, 2016, Ghaffarian Hoseini et al., 2013). Such a policy paradigm relies primarily on provision of electricity and enhanced access to internet, telephone and wireless networks.

In the case of Indian villages, Viswanadham and Vedula (2014) defined a Smart Village as "a bundle of services delivered to its residents and businesses in an effective and efficient manner" through a village ecosystem approach that brings all the services of the village and its providers and users on a single platform. Ragunathan and Sultana (2017) have articulated SMART as an acronym of a wider conceptualisation.

There are many such articulations of "smart" but at the core is the idea of technology as a tool for integrating the local with global and that this encourages entrepreneurial attitudes among residents to achieve self-sustainability.

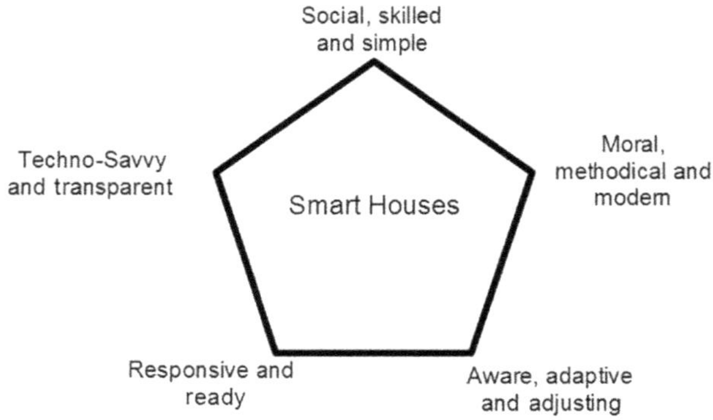

Figure 5.3 Smart houses in Smart Villages

The emphasis on technology determinism is pronounced in the vision of the Indian government where it has outlined 14 components in a list of parameters to assess success of rural development covering a variety of skills development and infrastructure services (Ragunathan & Sultana, 2017).

Critics of technology-led rural development argue that improvements in technology should be accompanied with the provision of good education, healthcare, clean water, sanitation and nutrition (Banda et al., 2007, Abdul Razak et al., 2013). In a Malaysian Smart Village project, Abdul Razak et al. (2013) suggest three phases for achieving sustainable rural development goals: social empowerment, developing the Smart Village ecosystem and economic empowerment.

Much of the discourse works at the level of the total environment or system level. It is surprising to see that houses and housing do not feature prominently in articulations of Smart Village. Ideas about smart houses have only recently begun to take shape. Ghaffarian Hoseini et al. (2013) suggest the basic concept focuses on two constituents: it has to be fully integrated with ambient intelligence environments, and it has to be based on interactions between users and environments. Figure 5.3 illustrates another conceptualisation of what constitutes a smart house. Although smart houses are in nascent stages, there is already a critique developing around sociocultural aspects of these houses as the emphasis on technology in creating a completely controlled environment takes away the interpersonal interactions between people and their natural settings (Abo-elazm & Ali, 2017). Many studies conclude that values and beliefs and integration of natural local life of the local region should be incorporated in the design of smart houses if they are to make a meaningful contribution towards sustainability.

Recommendations for best practice

Once based on agriculture as a primary occupation, rural economies are increasingly diversifying. The pace of transformation is even faster as new technologies are made available in rural environments. Some of the newer activities are rural tourism, cottage industries, start-ups, rural firms and holiday homes. These are creating demand for changes in predominant land-use patterns in rural areas. This also means that emerging forms of housing will be different from the traditional systems and livelihoods.

The application of recent approaches of Smart Village and smart houses represent a new milestone in the changing rural landscape. However, such approaches should also be sensitive to the defining characteristics of rural areas, including vernacular architecture through which most housing is produced. The intersection of new technologies with traditional modes of housing provisions should be addressed in a holistic manner and with due considerations to the sociocultural sustainability of housing. Chiu (2004) has offered a framework for examining concepts of social sustainability and cultural sustainability and argues that these two are interlinked and key dimensions of sustainable development. Chiu's framework explains social sustainability in terms of three interrelated elements: "social constraints limiting development; social pre-conditions determining distribution of resources and assets within and over generations; and maintenance or improvement of the well-being of people". These are development-oriented, environment-oriented and people-oriented respectively. Her definition of cultural framework builds on the work of Rapoport and includes aesthetic and artistic aspects (fine arts, music, popular culture, performing arts etc.), the cultivation of mind and spirit (knowledge, belief, religion and ideologies) and way of life. She highlights the importance of cultural factors in this framework as "cultural values and customs are often at play to set social norms" (Chiu, 2004).

While in general, "vernacular traditions are seen to be in a state of decline and are frequently looked down upon, abandoned, neglected or actively demolished" (Foruzanmehr & Vellinga, 2011), they can also provide an opportunity for reinvention with the incorporation of smart technologies that are increasingly becoming realities of contemporary rural societies. Including "social well-being" and "culture" in the subject of housing sustainability can lead to fruitful hybridisation (Chiu, 2004). In that sense, choices often driven by economic considerations and the need for effective use of environmental resources can really become smart.

We believe that smart integration can help overcome the negative perception of vernacular as something that does not reflect a contemporary and progressive outlook on life (Foruzanmehr & Vellinga, 2011). Our notion of smart lies in enhancing traditional wisdom while embracing automation and innovations in building materials and technologies. Laidin (2015) has demonstrated that ICT can help produce documentation for the preservation of culture, oral tradition and traditional knowledge. We advocate resisting the "historical entrapment of the vernacular" (Vellinga, 2006) and rather argue for encouraging smart houses and smart housing as a way to move forward in rural areas. The agenda of smart houses is promising in working towards "the harmony between the utilization of advanced technologies and local characteristics of region" and "the use of appropriate vernacular features of vernacular built environments as embodiment of smart living environments" (Ghaffarian Hoseini et al., 2013).

In rural societies, people "know how to make the shelters they need and want, and which will preserve the cultural forms that keep them secure" (Cromley, 2008). Historically, "the transmission, development, and amalgamation of building traditions" have acted as enablers for the development of an approach to architecture in such places (Vellinga, 2006). It is this confidence that leads us to suggest that newer cultural formations around smart technologies will emerge – how to ensure their connections with traditional systems is something that needs to be carefully catered to.

References

Abdul Razak, N., Abdul Malik, J. & Saeed, M. 2013. A development of smart village implementation plan for agriculture: A pioneer project in Malaysia. Proceedings of the 4th International Conference on Computing and Informatics, ICOCI 2013 Paper No. 28–30 August, Sarawak, Malaysia. Universiti Utara Malaysia.

Abo-elazm, F. M. & Ali, S. M. 2017. The concept of "Local Smart Architecture": An approach to appropriate local sustainable buildings. *International Journal of Cultural Heritage*, 2.

Abu-Ghazzeh, T. M. 1997. Vernacular architecture education in the Islamic society of Saudi Arabia: Towards the development of an authentic contemporary built environment. *Habitat International*, 21, 229–253.

Anna-Maria, V. 2009. Evaluation of a sustainable Greek vernacular settlement and its landscape: Architectural typology and building physics. *Building and Environment*, 44, 1095–1106.

Banda, K., Sarkar, R., Gopal, S., Govindarajan, J., Harijan, B. B., Jeyakumar, M. B., Mitta, P., Sadanala, M. E., Selwyn, T., Suresh, C. R., Thomas, V. A., Devadason, P., Kumar, R., Selvapandian, D., Kang, G. & Balraj, V. 2007. Water handling, sanitation and defecation practices in rural southern India: A knowledge, attitudes and practices study. *Transactions of the Royal Society of Tropical Medicine and Hygiene*, 101, 1124–1130.

Bański, J. & Wesołowska, M. 2010. Transformations in housing construction in rural areas of Poland's Lublin region—Influence on the spatial settlement structure and landscape aesthetics. *Landscape and Urban Planning*, 94, 116–126.

Bodach, S., Lang, W. & Hamhaber, J. 2014. Climate responsive building design strategies of vernacular architecture in Nepal. *Energy and Buildings*, 81, 227–242.

Chiu, R. L. 2004. Socio-cultural sustainability of housing: A conceptual exploration. *Housing, Theory and Society*, 21, 65–76.

Cromley, E. 2008. Cultural embeddedness in vernacular architecture. *Building Research & Information*, 36, 301–304.

Dili, A., Naseer, M. & Varghese, T. Z. 2010. Passive environment control system of Kerala vernacular residential architecture for a comfortable indoor environment: A qualitative and quantitative analyses. *Energy and Buildings*, 42, 917–927.

Dodo, Y. A., Ahmad, M. H., Dodo, M., Bashir, F. M. & Shika, S. A. 2014. Lessons from Sukur Vernacular Architecture: A Building Material Perspective. *Advanced Materials Research*, 207–210.

Foruzanmehr, A. & Vellinga, M. 2011. Vernacular architecture: Questions of comfort and practicability. *Building Research & Information*, 39, 274–285.

Fuentes, J. M. 2010. Methodological bases for documenting and reusing vernacular farm architecture. *Journal of Cultural Heritage*, 11, 119–129.

Gautam, A. 2008. *Climate responsive vernacular architecture: Jharkhand, India*. Manhattan, KS: Kansas State University.
Ghaffarian Hoseini, A. H., Dahlan, N. D., Berardi, U., Ghaffarian Hoseini, A. & Makaremi, N. 2013. The essence of future smart houses: From embedding ICT to adapting to sustainability principles. *Renewable and Sustainable Energy Reviews*, 24, 593–607.
Gkartzios, M. & Scott, M. 2010. Residential mobilities and house building in rural Ireland: evidence from three case studies. *Sociologia Ruralis*, 50, 64–84.
Holmes, J., Jones, B. & Heap, B. 2015. Smart villages. *Science*, 350(6259), 359.
Indraganti, M. 2010. Understanding the climate sensitive architecture of Marikal, a village in Telangana region in Andhra Pradesh, India. *Building and Environment*, 45, 2709–2722.
Joon, V., Chandra, A. & Bhattacharya, M. 2009. Household energy consumption pattern and socio-cultural dimensions associated with it: A case study of rural Haryana, India. *Biomass and Bioenergy*, 33, 1509–1512.
Kaza, N. 2010. Understanding the spectrum of residential energy consumption: A quantile regression. *Energy Policy*, 38, 6574–6585.
Kimura, K.-I. 1994. Vernacular technologies applied to modern architecture. *Renewable Energy*, 5, 900–907.
Kirk, J. 2001. Media and social change: The modernizing influences of television in rural India. *Media, Culture & Society*, 23, 147–169.
Kumar, A., Tewari, A., Shroff, G., Chittamuru, D., Kam, M. & Canny, J. 2010. An exploratory study of unsupervised mobile learning in rural India. *Proceedings of the SIGCHI Conference on Human Factors in Computing Systems*. Atlanta, Georgia, USA: ACM.
Laidin, A. Z. 2015. Smart villages – The Malaysian approach. *In*: Holmes, J. et al. (eds.), *Smart Villages: New thinking for off-grid communities worldwide*, 42.
Laxmi, V., Parikh, J., Karmakar, S. & Dabrase, P. 2003. Household energy, women's hardship and health impacts in rural Rajasthan, India: Need for sustainable energy solutions. *Energy for Sustainable Development*, 7, 50–68.
Lim, R. M. 2007. Cultural sustainability and development: Drukpa and Burman vernacular architecture. *In*: Zetter, R. & Butina Watson, G. (eds.), *Designing sustainable cities in the developing world*. Farnham: Ashgate, 21.
Lopez, S. L. 2010. The remittance house: Architecture of migration in rural Mexico. *Buildings & Landscapes: Journal of the Vernacular Architecture Forum*. Minneapolis: University of Minnesota Press, 33–52.
Lucon, O., Ürge-Vorsatz, D., Ahmed, A. Z., Akbari, H., Bertoldi, P., Cabeza, L. F., Eyre, N., GadgilL, A., Harvey, L. D. D., Jiang, Y., Liphoto, E., Mirasgedis, S., Murakami, S., Parikh, J., Pyke, C. & Vilariño, M. V. 2014. Buildings. *In*: Edenhofer, O., Pichs-Madruga, R., Sokona, Y., Farahani, E., Kadner, S., Seyboth, K., Adler, A., Baum, I., Brunner, S., Eickemeier, P., Kriemann, B., Savolainen, J., Schlömer, S., Stechow, C. V., Zwickel, T. & Minx, J. C. (eds.), *Climate change 2014: Mitigation of climate change*. Cambridge and New York: Contribution of Working Group III to the Fifth Assessment Report of the Intergovernmental Panel on Climate Change.
Maudlin, D. 2010. Crossing boundaries: Revisiting the thresholds of vernacular architecture. *Vernacular Architecture*, 41, 10–14.
Nepal, S. K. 2007. Tourism and rural settlements Nepal's Annapurna region. *Annals of Tourism Research*, 34, 855–875.
Nguyen, A.-T., Tran, Q.-B., Tran, D.-Q. & Reiter, S. 2011. An investigation on climate responsive design strategies of vernacular housing in Vietnam. *Building and Environment*, 46, 2088–2106.

Pauli, J. 2008. A house of one's own: Gender, migration, and residence in rural Mexico. *American Ethnologist*, 35, 171–187.

Ragunathan, M. V. & Sultana, M. F. Smart Cities & Smart Villages: A Mission to Transform the Nation. *In*: Ragunathan, M. V. & Sultana, M. F. (eds.), National Conference on Smart Cities and Smart Villages: A mission to transform the nation, 2017 Kodaikanal. *Shanlax International Journal of Management*.

Rapoport, A. 1987. On the cultural responsiveness of architecture. *Journal of Architectural Education*, 41, 10–15.

Rapoport, A. 1998. Using "Culture" in housing design. *Housing and Society*, 25, 1–20.

Sai, S. S. T. 2016. Smart Villages – Need of emerging India. *International Journal Of Innovative Research in Information Security*, 9.

Shastry, V., Mani, M. & Tenorio, R. 2014. Impacts of modern transitions on thermal comfort in vernacular dwellings in warm-humid climate of Sugganahalli (India). *Indoor and Built Environment*, 23, 543–564.

Singh, M. K., Mahapatra, S. & Atreya, S. 2009. Bioclimatism and vernacular architecture of north-east India. *Building and Environment*, 44, 878–888.

Singh, M. K., Mahapatra, S. & Atreya, S. 2011. Solar passive features in vernacular architecture of North-East India. *Solar Energy*, 85.

Tiwari, P. 2001. Housing and development objectives in India. *Habitat International*, 25, 229–253.

Upton, D. 1983. The power of things: Recent studies in American vernacular architecture. *American Quarterly*, 35, 262–279.

Vellinga, M. 2006. The inventiveness of tradition: Vernacular architecture and the future. *Perspectives in Vernacular Architecture*, 13, 115–128.

Viswanadham, N. & Vedula, S. 2014. *Design of Smart Villages. Computer science and automation*, Indian Institute of Science, Bangalore.

Yıldırım, M. & Turan, G. 2012. Sustainable development in historic areas: Adaptive re-use challenges in traditional houses in Sanliurfa, Turkey. *Habitat International*, 36, 493–503.

Zhai, Z. J. & Previtali, J. M. 2010. Ancient vernacular architecture: Characteristics categorization and energy performance evaluation. *Energy and Buildings*, 42, 357–365.

6 Energy

Introduction

The seventh UN Sustainable Development Goal is "Affordable and Clean Energy for all" (United Nations, 2017). Access to energy is no longer considered a convenience but a basic human right. Both the UN and the International Energy Agency estimate that worldwide one in five people lack access to electricity and three billion rely on wood, coal, charcoal or animal waste for cooking and heating. Access to electricity is a significant contributor to the urban-rural divide. For example, in India 96 per cent of urban residences have access to grid supplied electricity, while only 67 per cent in rural areas have access (Barman et al., 2017). The International Energy Agency reports that 84 per cent of people without access to electricity live in remote rural communities (Prinsloo et al., 2017). Palit and Chaurey (2011) point out that while the percentage of populations with access to electricity has increased, the total number of people without access has not decreased. There is a combination of reasons for this including the fact the global population has increased at the same rate as electrification, however, some areas that are connected to the grid may have had power turned off due to inability to meet demand.

Energy generation is the greatest emitter of greenhouse gases (GHG), accounting for 60 per cent of the total GHG emitted into the atmosphere. The accumulation of GHGs in the atmosphere is the primary cause of climate change. Some of the biggest impacts of climate change on rural communities are in Asia and sub-Saharan Africa. Ironically, people in these communities are the least likely to have access to energy themselves and are bearing the brunt of the impacts of climate change caused mainly by energy generating activities in more developed countries including the US and China while receiving none of the benefits associated with having a readily available supply of energy. This is one of the main reasons some scholars believe access to energy is intrinsically linked with issues of justice. Islar et al. (2017) defines energy justice as:

- a concept for philosophers and ethicists to better integrate distinct distributive and procedural justice concerns;
- an analytical tool for researchers striving to understand how values are presented in energy systems and/or energy related conflicts;
- a decision-making instrument that assists energy planners and consumers in making more informed energy choices.

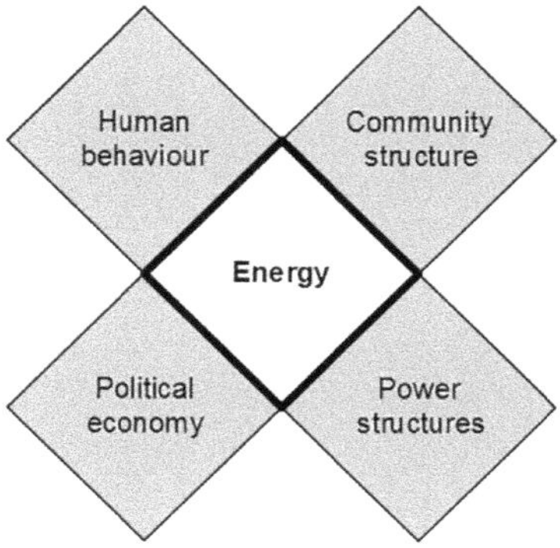

Figure 6.1 Relations between energy, human behaviour, community structure, political economy and power structures (adapted from Islar et al., 2017)

Islar et al. (2017) also highlights the need for energy studies to include social science perspectives as illustrated in Figure 6.1.

Sovacool and Dworkin (2015) identify eight components they believe are required to achieve energy justice as shown in Table 6.1.

Table 6.1 Eight components of energy justice

Component	*Description*
Availability	Connection to an energy supply that supports all the requirements
Affordability	All households can afford their energy bills
Due process	All stakeholders have opportunity to comment on policy decision-making and post-implementation grievances should be addressed via neutral arbitration
Good governance	The energy policy decision-making process should be transparent
Sustainability	Energy demands should be met without destroying the ability of the environment to provide for future generations
Intragenerational equity	All members of the population should have equal access to energy resources independent of social standing, gender etc.
Intergenerational equity	Distribution of energy resources should be equal between current and future generations
Responsibility	Governments are responsible for minimising environmental degradation, developed nations are responsible for the current pollution issues related to energy generation, the current generation is responsible for the welfare of future generations, and humans are responsible for the welfare of other species

Reference: Sovacool and Dworkin, 2015

Thus, the ultimate goal of rural electrification is to bring an adequate and reliable energy supply to all households, using an energy source that is safe, affordable, renewable and non-polluting. Although this might seem a little fantastical, there are examples of where this has been achieved. Costa Rica, for example, has an electrification rate of over 99 per cent, which is predominately provided by renewable energy sources. In fact, in 2015, they supplied all of their energy needs from renewable sources for 299 consecutive days, and beat this record in 2017, generating all of their energy needs from renewables for more than 300 days (Embury-Dennis, 2017). Of course, Costa Rica has some advantages over other countries such as an abundance of active volcanoes providing opportunities for geothermal energy. It is also a mountainous country with heavy rainfall creating optimal conditions for hydroelectricity generation. However, it does give hope that it is possible to generate a significant proportion of our energy supply from renewable sources.

In this chapter we explore the feasibility of achieving ubiquitous, rural electrification starting with a discussion of the benefits that access to clean and affordable electrification could bring to rural communities. We then discuss the different distribution options, from extending a national grid system versus microgrid solutions. Finally, different microgrid technological options for achieving the goals of energy access and justice, including energy storage, are reviewed.

Electricity

The benefits of electricity provision

Many households worldwide living without electricity rely on kerosene lamps for lighting. Estimates of the total number range from 500 million (Lam et al., 2012) to over a billion (WHO, 2016). The use of kerosene lamps has many disadvantages:

- It is very expensive, often taking up a significant portion of a household's income. From the consumer perspective the cost of electricity is much lower for example Gippner et al. (2012) report that in Nepal one household was paying 600 rupees per month for kerosene. Since connecting to electricity they are now paying only 100 rupees per month.
- It is not always readily available. Gippner et al. (2012) also report that some people had to travel up to three days away to purchase kerosene, thus on top of incurring the costs of purchasing kerosene people are also losing significant portions of their time that could be spent in income generating activities.
- There is risk of fires and explosions (Lam et al., 2012).
- Kerosene burning leads to the formation of pollutants due to incomplete combustion including carbon monoxide, nitrous oxides, sulphur dioxide and fine particulate matter. Its use in an indoor environment contributes to poor indoor air quality, which some health studies suggest can lead to respiratory diseases, including lung cancers (Lam et al., 2012).

- The quality of light from kerosene lamps is also very poor, making it difficult to perform normal evening tasks like studying or sewing. Thus, lack of access to electric lighting limits the amount of time people can spend performing tasks that could generate income, or the amount of time they can spend studying, hindering their ability to improve their quality of life through increased income and obtaining better education (United Nations, 2017).

While many rural electrification projects have focused on lighting provision, these have often fallen short of the community's requirements. Barman et al. (2017) investigated the failure of a rural electrification project in India and found that the main reason for the shortfall was most households also needed to charge a mobile phone. Even in non-electrified communities, many people have mobile phones and are travelling to a charging station, sometimes over long distances, and paying a fee that would be much higher than the cost of an electricity tariff. Electricity can also allow the use of radios, televisions and other forms of communication, increasing resident's ability to engage with the world. Therefore, prior to provision of electricity the requirements of households in the community must be established. At the University of Melbourne, we surveyed 570 households on the island of Majuli, Assam, India, a population consisting of 2000 households distributed in 37 rural villages. Figure 6.2 shows the

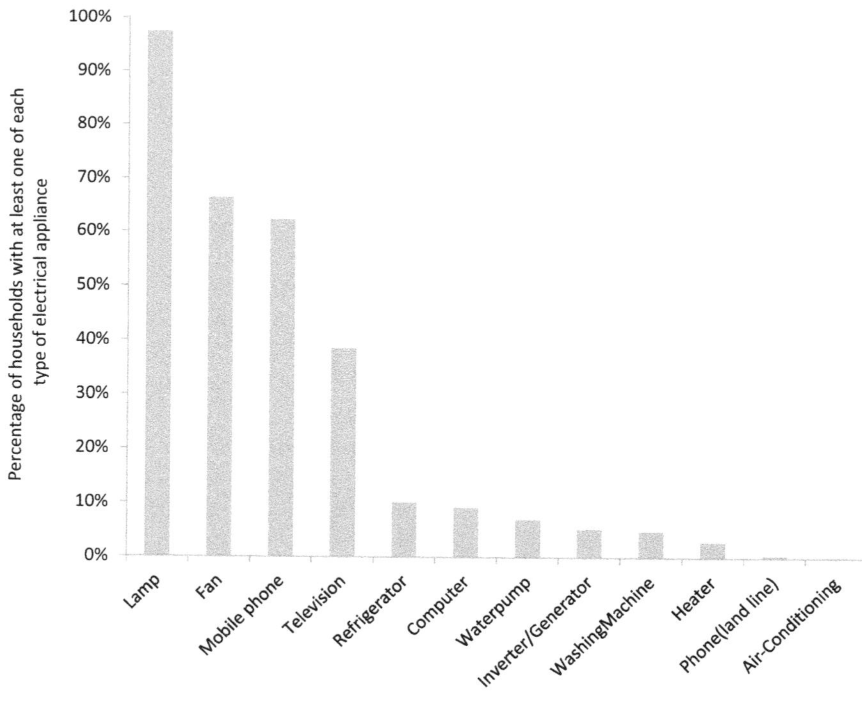

Figure 6.2 Percentage of households from Majuli Island with at least one of each type of electrical appliance

percentage of electricity connected households with each type of appliance. As can be seen, lamps are by far the most commonly owned product. However, fans and mobile phones are also widely used, both owned by over 60 per cent of households. Televisions are the next most commonly owned product, at around 40 per cent.

The provision of electricity has the potential to significantly improve the quality of life of a household's members by providing opportunities to increase income and educational attainment. Ray et al. (2016) compared three human development indicators between villages using raw biomass for heating and cooking, and kerosene lamps for lighting, to those using charcoal for heating and cooking and solar panels for electric lighting. The three indicators explored were education level based on school enrolment and literacy, healthcare level based on life expectancy, per capita GDP based on income level. The HDI of those using modern energy techniques was improved 16–18 per cent compared to the group using raw biomass and kerosene.

There are many ways that electricity can increase income generation. People can use the extended hours of light to pickle their home-grown vegetables, enabling them to create more produce for sale and reducing the risk of post-harvest losses. A tailor would be able to work for several more hours in the evening, increasing the workload they are able to take on. Similarly, households that sell their produce directly from home will be able to extend their opening hours. In many occupations, it could also mean a huge increase in productivity, for example the purchase of an electric sewing machine could substantially increase a tailor's income.

Agricultural activities can be electrified which can lead to a huge saving of time, especially for women who are often tasked with time-consuming manual labour. Rice is a good example as it is the third most commonly grown crop worldwide, and the most widely consumed staple. Figure 6.3 depicts an Assamese woman harvesting rise. Post-harvest, rice needs to be processed to remove the husk and bran layers from the rice kernel. The removal process is known as milling. Traditionally, rice has been milled by hand using a pestle and mortar, and this process is still practised in many small farming communities in developing economies. This process is extremely time-consuming and can lead to breakages of the rice kernel, reducing its market value. Often broken rice will be sold as brewer's rice (for making alcohol) or ground into rice flour or noodles. Establishing a mill in a village can allow farmers to drop off their rice here to be processed automatically, in approximately one hour. Legros et al. (2011) estimated that automated rice milling frees up 155 hours per year for women, and for men it would save 85 hours per year. The improved quality of the rice will allow the farmer to sell it for a higher price, and the extra time they have available could be used to generate income in other ways, or pursue education increasing their employability. The mill will require an operator and thus this also creates a job for another member of the community. Another important way that electrification could increase a farmer's income is allowing for the operation of a radio or television which would give farmers access to weather reports. As described in Chapter 4, knowing the weather can lead to increases in yields by indicating the most advantageous times for planting, adding fertilisers, irrigation etc.

Figure 6.3 Assam rice harvesting

Educational attainment can also be enhanced in multiple ways by electricity access (Legros et al., 2011). Villages with electricity are more appealing places to live, and therefore, are more likely to attract teachers to live in the area, increasing teacher/student ratios. The increased income of the average household, and reduced time required to perform income generating tasks, means children will not be required to spend as much time working in the family business, giving them more time to attend school. Similarly, families will have more income, allowing them to spend more on their children's education. The provision of good quality lighting during the evening will allow children more time to study, giving them greater learning opportunities. Schools will be able to operate computers and other electronic learning aids, increasing educational opportunities and ensuring children are prepared for the modern world.

Distribution options

There are two broad approaches to bringing electricity to rural communities. The first is to extend the main national electrical grid that carries power from sources to demand centres and individual consumers. The second is to provide "microgrid" solutions, where electricity is generated locally. The advantages of connection to the national grid are that sources and distribution systems are managed by few people. Therefore, when problems occur, or maintenance is required, there are experts available to manage these tasks. The average householder does not need to understand how the system works and can simply flick a switch in their household to obtain energy. Microgrid solutions can be more appropriate where the technical requirements or costs of extending the grid are prohibitive.

Mahapatra and Dasappa (2012) performed a life cycle cost analysis comparison of connecting to a national grid, to an islanded microgrid system with PV cells and an islanded microgrid with gasification plant. The formula below shows the determination of the life cycle cost of the microgrid systems (LCC_{RES}):

$$LCC_{RES} = \frac{CAPEX + OPEX + Component\ Replacement - Carbon\ benefit}{Capacity \times annual\ hours\ of\ operation \times lifetime}$$

This gives the LCC in money per kWh. While national grid extension lifecycle costs (LCC_{GE}) were calculated as:

$$LCC_{GE} = \frac{LCC_{Gen} + LCC_{Grid} \times Distance\ from\ powerplant}{Capacity \times annual\ hours\ of\ operation \times lifetime}$$

The grid extension calculation results were dominated by *the distance from power plant* term indicating this is the most significant factor when deciding between grid extension and islanded microgrid system. This makes sense as transmission lines experience losses, and the further distance required the greater loss (Wirfs-Brock, 2015). In between populated areas electricity is transmitted on higher voltage lines, which reduces losses to around 2 per cent. Due to the high voltage travelling along these lines, they are extremely dangerous, therefore closer to populated areas for distribution of electricity the lines run a much lower voltage, which means they experience greater losses of around 4 per cent. The fewer people in the populated region, the greater the loss of electricity.

While this calculation provides a good indication of the cost differential between national grid extensions versus microgrid solutions, there are several other factors to consider. Transmission losses are also impacted by the weather and by spikes in energy use (Wirfs-Brock, 2015). Transmission lines are also at risk of "non-technical" losses in remote locations – where people tap into the lines without paying. The costs of extending the grid could also be higher depending on the terrain of the region. Transmissions lines could also spoil landscapes and disturb ecosystems which could be detrimental to the livelihoods of some people, for example those related to ecotourism.

Perhaps the most significant factor the above calculation fails to consider is reliability. In developed countries we have come to rely heavily on a constant supply of electricity. For example, in the 2003 blackouts in New York 90 deaths occurred as a direct result of electricity loss, from people on life support machines in the hospital, inability of ambulances to travel and inability to obtain prescription medications (Bird & Hotaling, 2017). In developing economies on the other hand, power outages are frequent due to the inability of the grid system to withstand peak loads. Small village communities are a low priority and therefore are the most likely to be cut off from the grid when the demand exceeds supply. In fact, there are households located in grid connected areas that choose not to connect. For example, in a recent survey on the island of Majuli, Assam, India which has ubiquitous electricity provision, approximately 20 per cent of households reported they used no electricity. The main reason is that households cannot afford both connection to the grid and a back-up supply of kerosene to use during blackouts. Thus, they perceive that the grid electricity services are not worth the money and would rather purchase reliable kerosene.

Microgrid systems are a very viable alternative to national grid extension. Currently, over 50 per cent of installed microgrids are in remote locations (Prinsloo et al., 2017). These systems can vary in scale from a small photovoltaic (PV) cell providing power to an individual household, to a village-wide system providing hundreds of kilowatts, or even a few megawatts. While this gives greater control over energy provision to the user, ensuring they have electricity when they need it most, at the same time it gives them responsibility of the operation and maintenance of the system, and if this is poorly understood, as would be the case for a lot of uneducated poor people in villages, then they could misuse the system, leading to lower levels of electricity than expected, or even damage it. Usually, once damaged there is little option for repair. Palit and Chaurey (2011) explored microgrid versus national grid options for South Asia and found that beyond the technical feasibility of extending the national grid, the choice also needs to consider resource availability, technical skills, socio-economics and governance setting.

A lot of microgrids have focused on renewable technologies, as this removes the operational costs associated with the purchase of fuels as well as having a lower burden on the environment. The price of fossil fuels is subject to wide, unpredictable variations (Williams et al., 2018) which make them risky for a small low-income community. Many of the existing microgrid systems currently in use have been NGO sponsored. The nature of these organisations is generally to align with Sustainable Development Goals further motivating the choice of renewable technologies. Early on a lot of these systems failed, however, developers have learned from these mistakes and there are now many successful up and running renewable energy based microgrid systems. It is not feasible to expect NGOs to continue establishing electrification; there must be a shift to private investors taking over. While the technical feasibility of renewable electricity microgrids is now proven, they must also prove their commercial viability. The first step is whether households will be willing and able to pay their tariffs. Duby and Engelmeier (2017) report that microgrid tariffs operate at a lower cost for an average household than the cost of kerosene they were previously using. Therefore, this should not be an issue, as long as the supply of electricity is reliable. This is another argument against fossil fuel based microgrid systems. Rural electrification projects are often introduced with a fixed price tariff to ensure affordability. Thus, if the system were fossil fuel based, any sudden increases in fuel price would be worn by the electricity company. Alternatively, introducing variable priced tariffs, which puts the burden of fuel prices on the user risks households being unable to afford the bills (Williams et al., 2018). It is also very likely the private investment will be international, leaving investors vulnerable to fluctuations in the exchange rate.

Another hindrance to the commerciality of microgrid systems is insecurity among potential investors about the government's expansion plans for the national grid. In many regions, these plans are not transparent, or if they are available the national grid connection may be a few years off. However, to an investor in a microgrid system this is a big risk as the payback period can often be on the scale of five to ten years, so if the national grid comes along offering electricity at a cheaper price and many of their customers switch, they will lose out on their investment

(Duby & Engelmeier, 2017). Alternatively, the national grid could incorporate the microgrid system during expansion, increasing diversity of fuel supply, and thereby reducing blackouts and increasing reliability (Bird & Hotaling, 2017). This solution could work for everybody; the national grid increases its electricity supply, improving its ability to meet demand, while the microgrid can sell excess energy to the national grid avoiding wasted energy generation where demand is below supply. This type of agreement will depend on the political climate of the country. Control of the provision of power is a source of political capital, and it may be difficult to convince some governments to free up the system, not just allowing, but supporting commercial microgrid systems to operate in a profitable way (Duby & Engelmeier, 2017).

There may be difficulties encouraging households to ditch kerosene and connect to the grid, if they have had a previous negative experience. Therefore, proving the reliability of the electricity supply will be an essential step for private investors. It will also be important to ensure demand in the area is sufficient to absorb the amount of electricity generated (Williams et al., 2018). An important way to do this is by supporting commercial activities, which generally have a higher electricity demand than residential properties. While the provision of electricity opens up the potential for establishing many types of new commercial activities, residents of rural communities may not be able to see this potential and require training to be made aware of it. In some Nepalese villages, electricity provision was coupled with a training programme highlighting potential business opportunities for the residents to partake in (Legros et al., 2011). A post-implementation comparison between one of these villages, and one that was provided with electricity only found that three times as many new businesses had been established in the region with the training programme, compared with other similar regions that were provided with electricity only. The success of this type of training also needs to be culturally sensitive. In Vanuatu electricity was provided along with training to encourage residents to establish income generating activities (Fischer & Pigneri, 2011). However, the local people in Vanuatu are not motivated by money. They are satisfied if their income provides them with the basics, a roof over their head and enough to eat. They are not interested in purchasing products surplus to this and therefore do not feel the need to increase their income beyond this.

Another issue facing new commercial activities in a rural community is obtaining start-up capital, and funding to allow them time to become established. Electricity suppliers will need to work with small commercial enterprises to help them grow before they become reliable customers. One potential way to do this is to provide pay-per-use electric services such as leasing power tools or refrigeration equipment. They could also offer services for residential customers who are wary of connecting to the grid such as a mobile phone charging station, or an internet café (Duby & Engelmeier, 2017).

Electricity demand often grows over time and another important factor to make a microgrid system viable is incorporating the potential for the system to expand. Failure to correctly anticipate the long-term electricity requirements of a project is one of the main reasons many microgrid electrification projects have failed (Akinyele & Rayudu, 2016, Williams et al., 2018). As described above, electrification has the potential to significantly increase a household's income. This is likely to lead to the accumulation of more electrical appliances. For example, a tailor provided with electricity for four hours of light in the evening will be able to work for four more hours (even if they had a kerosene lamp, the quality of light would not be adequate to see the fine detail of stitching). With this increased income, the tailor may decide to purchase an electric sewing machine to further increase productivity, and thus their income. Any microgrid electrification project must therefore incorporate an annual increase in electrical demand to remain fit for purpose.

Once a microgrid is established with a mix of commercial and residential customers, there will no doubt be fluctuations in demand, just as with the national grid system. Therefore, the distribution of electricity from a centralised microgrid system will need to incorporate an energy management system (EMS) to cope with these fluctuations in supply and demand (Prinsloo et al., 2017). There are four broad options for EMSs. Top-down switching is a centrally controlled EMS that allows a central management to switch electricity between users when required. The second is a price reaction system, so as demand increases so does the price of electricity. In this system the price for operating a particular appliance at a given time will be communicated to the user allowing them to choose whether or not to operate that appliance at that time. Both of these EMSs have a one-way communication system and react to demand in real time. This means that there is little ability to forecast electricity demand and improve availability. For example, if the cost is frequently too high to operate appliances users might become annoyed by the system and opt out, going back to using kerosene lamps. The other two options involve two-way communication, so the system is recording both demand and supply, allowing better forecasting of future electricity requirements. The first of these options is a centrally controlled system, but this may not be most appropriate in a small rural community. Prinsloo et al. (2017) argue that the most appropriate system in this case is transaction control. This basically operates like an auction, where users set a limit on the amount they are willing to pay to operate each appliance, and suppliers can set a price they are willing to sell electricity for. The system is thus collecting data from both users and suppliers and over time will be able to more accurately project electricity supply and demand.

Microgrid technology options

There are many choices for technology of a microgrid system. While diesel generators are always an option, the fluctuating costs of fossil fuels, and high pollution emissions of these systems make them unfavourable. Here, we only consider renewable microgrid options. The best option will be highly location dependent as many of these are related to either weather conditions or natural landscapes.

Solar

Solar microgrids are mainly formed of photovoltaic (PV) cells as depicted in Figure 6.4. It is currently the most common microgrid solution. Solar is theoretically the ultimate solution to provision of electricity. It's renewable, there is no need to purchase expensive fuels for system operation, and it doesn't emit any pollution. Yet, reading through the literature, there are many tales of solar system failures (Barman et al., 2017). One of the main reasons is that experts were brought in from outside to install the system, and no or inadequate training on how the system works was provided to the local people. Therefore, when the system failed there was no one available to carry out repairs. Thus, accompanying installation with a training programme for local people on basic maintenance is essential for this to be a viable sustainable solution. This training must include some level of follow-up checks to ensure the training material has really been understood.

The main limitation of solar electricity is that there is no provision for expansion. A solar panel can produce a set amount of energy, if it is exposed to full sunlight for a set number of hours per day. To increase energy generation, the only option is to purchase more PV cells. The ability of the system to absorb an adequate amount of sunlight can also be an issue depending on the mean daily solar radiation, and any potential shading that may be surrounding the location of the cell. In regions close to the equator solar is most appropriate because these areas tend to receive consistent amounts of sunlight year round. Other regions, with more diverse temperature variations may not receive enough sunlight during the winter months to operate the system effectively. Solar panels are also vulnerable to system losses in three ways (Akinyele & Rayudu, 2016):

a) They are sensitive to ambient temperature. As the temperature of the array increases, the voltage, overall power and efficiency are all reduced. This sensitivity of PV systems to temperature is often overlooked in economic and technical feasibility assessments. However, many developing economies, where these systems are being implemented, are subject to extremely high temperatures, and therefore, these assessments fail to realistically project the potential of PV systems.
b) The incomplete utilisation of the solar irradiation due to shading, dust collecting on the cells, or soiling.
c) Balance of system losses due to inefficiencies or losses of inverter, wiring losses, mismatch etc.

An example of a failed solar initiative is in Assam, India where the government provided thousands of households with solar cells. Barman et al. (2017) surveyed 544 of these households and found that only 28.9 per cent were functioning properly, 62.3 per cent were operating with minor faults, 1.6 per cent operating with major faults and 7.2 per cent were non-functional. The main problem identified was that the people did not understand how the technology worked and had accidently damaged the equipment by misusing them. The cells were designed to provide energy for two lights for four hours a night, however, many of the people owned mobile phones, which they needed to charge. The programme design phase failed to consult with the residents to determine their electrical requirements before implementation, and thus it was not fit for purpose. Another issue

Figure 6.4 Solar power in north west India

was that the cells were not fixed in place at the household but supplied loose. The household members were left to position them, and many were placed in the shade or other unsuitable location, meaning that they were not fully charging, and therefore not able to provide the prescribed per day electricity supply. This led people to think they were not working properly and ultimately damaging them while trying to carry out repairs. A final issue, because the cells weren't "fixed" they were vulnerable to theft and so many people were keeping them indoors overnight. The constant moving of the cells meant they were rarely positioned appropriately for maximising energy absorption from the sun.

While some solar projects have failed, there are also many success stories, where some of the issues highlighted above were overcome. Azuri Technologies' PayGo system (Azuri Technologies Ltd, 2017) provides solar off-grid solutions to many countries in sub-Saharan Africa. Their model is to charge a one-off installation fee, followed by a pay-as-you-go top-up system. The system provides enough electricity for a light globe for eight hours per day, and to charge a mobile phone. The weekly top-up cost has been calculated to ensure that it is less than the weekly cost families were previously spending on kerosene. A survey of PayGo customers in Kenya found that people were also saving money, as previously they had been taking their mobile phones to a charging station which cost more than the PayGo system.

Nasir et al. (2018) propose that a microgrid system could be built up over time by a set of nanogrid systems. In this instance a nanogrid system consists of a solar panel placed at an individual household, with storage that can be used directly by that house. Any surplus electricity generated at the household can be fed into the microgrid system for use elsewhere. Once established, the system allows for easy expansion. The system could theoretically be established with a solar panel and a single household. Other households can easily be added in later, increasing the system capacity. This is a good approach as many rural communities have been victim to failed microgrid systems and are wary of buying into a new scheme. The cost of them participating in the scheme is likely to mean they forgo purchasing kerosene or other fuels, and if the scheme fails they will be left with no energy. If the scheme can be established with only a few houses participating, then other households can opt to buy in later, when they are more confident that the system will provide a reliable source of electricity.

Hydro

The operation of a hydroelectricity plant is considered environmentally friendly, as it does not emit greenhouse gases and avoids the use of fossil fuels for energy generation. However, the construction of hydroelectric dams can divert precious water resources away from other users. The construction of dams can sometimes require flooding of large areas, displacing indigenous tribes, and threatening biodiversity. The HidroAysén project in Chile's Patagonia region would have flooded 59 km^2 of pristine wilderness and displaced millions of indigenous people. Biologically, Patagonia is an important area of biodiversity as well as outstanding natural beauty (Maxwell, 2016). The diversion of water from naturally flowing rivers also threatened the country's drinking water supplies, and fishing industries. Therefore, this proposal was eventually rejected after a long campaign by environmentalists from both within and outside of Chile. In Chapter 8, we will explore the relationship between the different services and industries that are competing for water resources and ways to distribute water resources fairly.

Microhydro plants can provide a more positive solution, as they only require a small amount of water. The most recent designs ensure fish cannot enter the system, meaning they do not pose a threat to aquatic fauna. Microhydro plants typically range in capacity from 50 kW to 3 MW (Palit & Chaurey, 2011). In Nepal, the mountainous terrain makes it technically unfeasible to extend the national grid to many highland villages, but the potential for hydroenergy is huge. In 1996 a programme to establish microhydro plants was initiated (Gippner et al , 2012), which by 2012 had installed 295 microhydro plants, with plans for another 106 underway. A post-implementation assessment of the impact of the microhydro systems had very positive results. Financially, households were on average 150 USD per year better off due to the combination of savings associated with not having to purchase kerosene, and the ability to generate additional income. There is further potential for households to increase their income by 912 USD, however, household members would need education to understand where the potential for increasing their incomes lies, and financing to help these businesses get started.

Wind turbines

Wind turbines, as illustrated in Figure 6.5 and Figure 6.6, have a longer track record of use than other renewable sources. Between 1980 and 2010 it is reported that 400,000 small to medium wind turbines have been built in China. The technology can offer a cheaper per kWh electricity source, over their lifetime. Yet they do not seem to be the favoured option in many developing economies when it

Figure 6.5 Wind turbines in Jaisalmer, India

Figure 6.6 House with wind turbine in Jaisalmer, India

comes to options for rural electrification (Wiemann, 2013). The alliance for rural electrification identified the main barriers to wind energy use as lack of awareness, limited production, difficulty in determining the most appropriate location and poor installation leading to system failures. Another difficulty highlighted in response to this article is the fact that wind turbines need to be imported to most low- and middle-income countries as they are mainly manufactured in the US and UK. The cost of importing these is often prohibitively expensive and is another barrier to their widespread use.

Bioenergy

The agricultural nature of the majority of income generating activities in rural areas means there is a large amount of organic waste to deal with. This could be considered an environmental problem, as its inappropriate storage and disposal poses a serious risk to human health and can lead to significant GHG emissions. However, the methane generation potential of this type of waste can provide a valuable source of renewable energy. As stated in the introduction, many rural households use raw biomass burning for cooking and heating. The open burning of biomass indoors is often romanticised as a natural and therefore safe way to provide energy. Organic matter reacts with oxygen producing heat, plus water and carbon dioxide:

$$CH_2O + O_2 \rightarrow CO_2 + H_2O + heat$$

$$Wood + natural\ oxygen\ in\ the\ air \rightarrow carbon\ dioxide + water + heat$$

However, the combustion process is incomplete and releases other chemicals into the air including particulate matter and carbon monoxide as well as smaller concentrations of other dangerous pollutants such as hydrocarbons, benzene and formaldehyde (Pennise, 2005). The health effects of particulate matter have been studied extensively and shown to lead to lung disease. Carbon monoxide is an odourless and colourless gas which has a similar mass to O_2 but reacts 250 times more readily with haemoglobin in the bloodstream. Thus, inhalation of CO deprives the body of oxygen leading to suffocation. Therefore, open burning of raw biomass is hazardous to health, and can also lead to fires and explosions.

Although it is possible to obtain electric cooking and heating appliances, gas-based fuels for these purposes are usually more efficient. The California Energy Commission shows that cooking with an electric stove uses around three times more energy than a gas stove (Watson, 2018). Gas cooking also gives the user greater control over the level of heat. We also discovered that while the majority of households on the island of Majuli use more than one fuel for cooking, only 19 households used electricity for cooking, and only one of these used electricity as their sole fuel for cooking. Similarly, only 33 houses had access to gas pipeline connection, but only five of these used this connection as their sole method for cooking.

Gasification is a safer and more efficient way of generating energy from biomass. In gasification, the biomass feedstocks are heated with a limited supply of oxygen to prevent combustion. The gasification process goes through several stages as the temperatures climb until they reach around 973 K. At this temperature, the organic material is converted to a mix of three gases, $CO + H_2 + CO_2$, known as syngas and is the fuel that can then be burnt directly to produce energy for heating and/or cooking, or as part of a generator to produce electricity. Biomass gasifier microgrids typically produce between 10 kW to 500 kW (Palit & Chaurey, 2011). In India the use of gasifiers has been a negative experience. One of the main reasons, the gasifier system was not provided with a suitable engine, but rather one adapted from an old diesel generator, which led to substantial capacity derating (Palit & Chaurey, 2011). The by-product of gasification is known as biochar, and is a carbon rich substance that can be added to soils to improve structure (All Power Labs, 2018).

Anaerobic digestion (AD) is another way to exploit the methane potential of organic waste, and is discussed further in Chapter 7. One of the advantages of AD over gasification is that it leaves a nutrient rich sludge, which can be applied to soils to provide fertilisation. The warm temperatures reached during the digestion process are sufficient to sterilise the waste, preventing the application of harmful pathogens to soils. Although AD theoretically sounds ideal, many AD projects fail. In China an assessment of the 18 million digesters in operation found that less than 60 per cent were operating effectively. In Nepal, 260,000 biogas facilities were established between 1992 and 2010 with funding from the clean development mechanism (Islar et al., 2017). An assessment of a sample of these projects found that only around 53 per cent were operating optimally (Cheng et al., 2014). The main reason for system failures were leaking pipes due to lack of maintenance.

One of the main drawbacks of gasifiers and AD is the technical skill required for system operation and maintenance. Fischer and Pigneri (2011) explored the feasibility of introducing gasification for rural electrification in Vanuatu, where biomass feedstock is in abundance, and found that lack of technical skill could limit the potential of the technology. Preparation of the feedstock for example involves growing short rotation feedstocks, including fertilisation and harvesting. Then drying and sawing the wood into the correct shape to suit the gasifier are all highly technical skills before you even start using the gasifier. Cheng et al. (2014) found lack of system maintenance and unavailability of skilled workers was the main reasons for systems failing to operate optimally.

While it may seem there is an inexhaustible supply of organic waste in rural areas, as energy requirements of the village expand over time, waste may no longer provide an adequate amount of biomass feedstock. Determining the feasibility of a biogas system based entirely on waste feedstocks is confounded by the fact that many areas keep very poor or no records of waste generation. Therefore, it will be difficult to determine how much feedstock is available. Where waste biomass feedstocks are inadequate, people will use other biomass materials, potentially destroying forests or other important areas of biodiversity and conservation. Various tools have been developed to help inform the decision-making process in relation to biomass energy systems. Singh et al. (2008) developed a GIS based tool to assess the potential biomass feedstock in a given area to inform decision-making on the feasibility of a biomass based energy system. Siemons (2001) explored the economic viability of a gasifier system compared to a diesel power generator, using charcoal or wood. The results were dictated by the local price of diesel, but generally the gasifier systems were more expensive, especially when using wood instead of charcoal as the feedstock.

Setting aside land to grow crops specifically for biomass feedstock could be a more appropriate method to maintain control over feedstock supplies. Growing biomass fuels could theoretically provide rural communities with another potential source of income; however, it also increases the risks of forests and other important areas of biodiversity being converted to agricultural land.

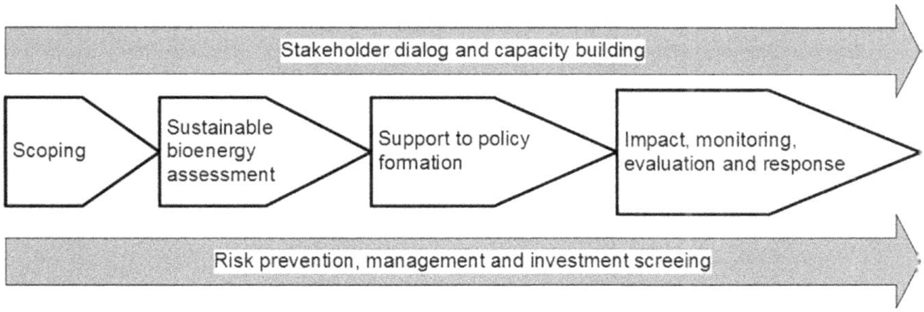

Figure 6.7 Bioenergy and food security approach (adapted from FAO, 2014)

Concerns have also been raised that setting aside arable land to grow crops for biomass fuels could compromise food security The Food and Agriculture Organization of the United Nations has developed a guide in response to this dilemma called the "Bioenergy and food security approach" as illustrated in Figure 6.7 (FAO, 2014).

Hybrid systems

From the discussion above it is clear that all of the renewable microgrid options suffer from limitations. Hybrid microgrid systems, which combine two or more technologies, have been shown to overcome these limitations. For example, López-González et al. (2017) compared various energy systems in Peru and found that a hybrid system minimised the risk of power outages that are frequent with single source electricity systems. Hybrid systems also often combine a renewable technology with a more traditional fossil fuel burning system that operates as a back-up when supply from the renewable system fails to meet demand (Siddaiah & Saini, 2016). Wind energy is especially adaptable to combination systems (Wiemann, 2013).

Kanase-Patil et al. (2010) compared four options for a hybrid microgrid system, as described in Table 6.2. They determined the scenario combinations based on an assessment of the abundance of the various natural energy sources in the area. They found scenario 4 to be the most feasible and sustainable.

Singh et al. (2016) optimised a wind-solar-biomass hybrid islanded microgrid system. The system was found to be affordable, if the optimum size of each fuel component is determined. The system incorporated a battery, which was the most expensive component; solar was the next greatest cost at 43 per cent, which also provided the greatest amount of energy supply. Wind was the next most expensive component at 6 per cent, and second highest energy supply. The biomass system was the cheapest system component at 4.5 per cent, but really only acted as back-up for the solar system during the darker winter months.

Table 6.2 Hybrid fuel scenarios

Scenario	*Fuel combination*
1	Hydro, biomass (from waste), solar
2	Hydro, biomass (from waste), solar, wind
3	Hydro, biomass (from dedicated plantation), solar
4	Hydro, biomass (from dedicated plantation), solar, wind

Reference: Kanase-Patil et al., 2010

Lai et al. (2017) assessed a combined solar – AD system for a village in Kenya. Due to its proximity to the equator Kenya is ideal for solar energy as it has relatively consistent daily solar radiation year-round. Kenya also has a large agricultural sector generating substantial amounts of organic waste to feed the AD system. Solar acts as the primary energy source, while AD makes up the difference when the solar system cannot meet demand. The system also incorporates electrical storage to supply energy when demand exceeds both the solar and AD supply.

Electrical energy storage

For energy supply to meet demand, it must be available when needed. Unfortunately, electricity demand varies by time of day and by region (Noguchi, Personal Communication). For example, solar power is generated during the day when people are out working in the fields, while they require electricity for lighting when they return home after sunset. Therefore, to be feasible there must be a way to store the energy generated by the sun during the day, so it is available in the evening when it is needed. A similar case can be made for other forms of renewables. The basic principle behind electrical energy storage (EES) is that surplus energy generated throughout the day is stored until demand exceeds generation, at which point the EES discharges. Incorporating EES into a hybrid microgrid system can provide the ultimate solution to reliable electricity supply. In these types of combined systems the EES can work in three ways: zero charge where only the solar source charges the EES; full cycle, where all the different fuel sources charge the EES so that it is always retained at full capacity; predictive control where the EES is maintained at a storage level that responds to the demand (Lai et al., 2017). The third strategy is considered the most advantageous as it avoids wasting energy generation. For example, in a combined solar-biogas system, where solar is the primary energy source and biogas acts as a back-up, the solar component will be able to charge the battery to the solar panels rated energy potential, then the biogas system will kick in to top up the battery charge. However, if the energy demand is less than the fuel charge of the EES, it will not be used up before the sun comes again the next day, and the biogas generation to top up the EES will have been a waste of fuel. Projecting the energy requirements of the system users will allow charging of the EES to the level required avoiding such waste. This is known as the State of Charge (SoC) and is similar to the fuel gauge in a petrol-powered car.

Energy storage can take various forms, which vary depending on the type of renewable resource. Pumped-storage in a hydroelectric plant is the most common method currently accounting for around 96 per cent of EES worldwide (World Energy Council, 2018). It involves having a second reservoir at a higher elevation than the main reservoir of the plant. When water flow in the main reservoir is low, the water from the higher reservoir pours down into the lower reservoir through turbines generating electricity. When the lower reservoir is full, excess water can be pumped into the higher reservoir, using only a small amount of electricity, ready to back the lower reservoir again when needed. This pumping is usually provided by the hydroelectric source itself, but could theoretically be fuelled by any source such as wind or solar, making it a potential storage system for these as well (Arnold, 2014). In a hybrid system, hydroelectric output can be adjusted up or down, corresponding to periods of high or low output from other sources in the hybrid system, making it a form of EES, without the requirement of pumped storage.

Batteries, which store energy in the form of chemicals, are the next most common form of energy storage. Energy is released through chemical changes to do electrical work on demand. Although they could theoretically be charged by any energy source, currently they are mostly used with solar. There are various types of batteries. Here we consider the two most common types currently used with solar PV cells: lead acid and lithium ion (Li-ion).

Lead acid batteries are the oldest form of rechargeable battery, as well as being the most widely used and cheapest on the market today (O'Connor, 2017). Li-ion batteries are a more recent and expensive technology. However, their current purchase price is a reflection of their recent emergence and will almost certainly decrease over time with price learning. While the purchase price of lithium ion batteries is much higher than lead acid, the "per kWh" lifetime costs of the two technologies are much more comparable, because Li-ion batteries have a substantially longer lifetime than lead acid.

Li-ion batteries are also more resilient than lead acid in several ways:

- They can withstand higher temperatures. The lifetime of lead acid batteries will be adversely affected at ambient temperature in excess of 298 K, while Li-ion batteries can withstand temperatures of up to 318 K. Therefore, Li-ion batteries are more suitable for warmer climates such as south Asia and sub-Saharan Africa. Exposing a lead acid battery to this climate would significantly reduce its lifetime, causing it to need replacing far more often than manufacture recommended, and thereby driving up the expense of the whole microgrid system (Duby & Engelmeier, 2017).
- Li-ion batteries are less sensitive to discharge. The lifetime of lead acid batteries will be adversely affected if they are discharged by more than 50 per

cent of their State of Charge, while Li-ion batteries can withstand discharge of up to 80 per cent. This will make Li-ion batteries particularly more suitable for a hybrid system where the SoC is maintained at a rate that reflects usage, avoiding wasting energy charging the battery to 100 per cent when this amount of energy exceeds demand.

- Li-ion batteries have a significantly higher energy density, making them smaller and lighter than lead acid batteries. This will make them easier to transport and store.

There are two other EES technologies that are less widely used. Energy from any source could be used in a hydrogen generator to electrolyse water, storing the resulting hydrogen, then converting it back to electricity when needed (Arnold, 2014). Similarly, compressed air can be stored in large above ground tanks or underground caverns. The air can then be used through direct expansion into a compressed air motor or injected in an internal combustion turbine where it is burnt with fuel to provide mechanical energy to a generator.

Recommendations for best practice

Setting up microgrid systems is recommended in communities that are not connected to the national grid. Even where national grid access is in the pipeline, it is recommended to set up a microgrid system, that can be incorporated into the national grid later to increase diversity of supply thereby improving system reliability.

The financing of microgrids in remote areas should start to move away from reliance on NGOs and incorporate private investment. Proving the financial viability of these systems is therefore a very important step.

The choice of microgrid technology will be location specific and careful assessment must be carried out to establish the energy generating potential of each option.

It is recommended that microgrids are established at the village level rather than household level, to decrease the number of people required to understand operation and maintenance of the system. Effective training in operation and maintenance must be provided at the time of implementation to ensure reliable long-term operation.

It is also recommended that the system contain more than one energy input where possible to reduce the risk of blackouts, even if this means a diesel generator may be required as a back-up in some locations.

Electrical energy storage should be operated with a predictive control strategy, where the charge level is somewhere between zero and 100 per cent, optimising the system and minimising energy losses.

The system should have the capacity to expand over time, as the increase in electricity will lead to an increase in income generating opportunities for many people, which in turn will lead to an increase in their electricity demands. This continued increase in income generation and increase in electricity demand is an important path towards alleviating poverty and achieving equality for remote rural communities. Thus, to be sustainable the system must be able to grow in response to this demand.

Finally, members of the community must be trained to manage and maintain the microgrid system as this is the only way the community will become self-sufficient. There are many stories of charitable organisations building a renewable microgrid system for a remote rural community that breaks down after a few months of operation, leaving the community no better off than they were before. This situation is avoided by training locals to maintain the systems. Often the systems break because of misuse due to the lack of understanding on the way the system works. Imparting this knowledge is crucial to the continued welfare of the community.

References

Akinyele, D. O. & Rayudu, R. K. 2016. Techno-economic and life cycle environmental performance analyses of a solar photovoltaic microgrid system for developing countries *Energy*, 109, 160–179.

All Power Labs 2018. *How gasification works* [Online]. Berkeley, California. Available: www.allpowerlabs.com/gasification-explained [Accessed 13 March 2018].

Arnold, B. 2014. *Collecting and storing energy from wind turbines* [Online]. Available: www.azocleantech.com/article.aspx?ArticleID=488 [Accessed 13 March 2018].

Azuri Technologies Ltd 2017. *Paygo solar power* [Online]. Cambridge, UK. Available: www.azuri-technologies.com/what-we-do [Accessed 21 February 2018].

Barman, M., Mahapatra, S., Palit, D. & Chaudhury, M. K. 2017. Performance and impact evaluation of solar home lighting systems on the rural livelihood in Assam, India. *Energy for Sustainable Development*, 38, 10–20.

Bird, S. & Hotaling, C. 2017. Multi-stakeholder microgrids for resilience and sustainability. *Environmental Hazards*, 16, 116–132.

Cheng, S., LI, Z., Mang, H.-P., Neupane, K., Wauthelet, M. & Huba, E.-M. 2014. Application of fault tree approach for technical assessment of small-sized biogas systems in Nepal. *Applied Energy*, 113, 1372–1381.

Duby, S. & Engelmeier, T. 2017. Kenya: The world's microgrid lab. Munich, Germany: TFE Consulting.

Embury-Dennis, T. 2017. Costa Rica's electricity generated by renewable energy for 300 days in 2017. *Independent*, 22 November.

FAO 2014. FAO's BEFS: Bioenergy and food security approach. FAO.

Fischer, B. & Pigneri, A. 2011. Potential for electrification from biomass gasification in Vanuatu. *Energy*, 36, 1640–1651.

Gippner, O., Dhakal, S. & Sovacool, B. K. 2012. Microhydro electrification and climate change adaptation in Nepal: Socioeconomic lessons from the Rural Energy Development Program (REDP). *Mitigation and Adaptation Strategies for Global Change*, 18, 407–427.

Islar, M., Brogaard, S. & Lemberg-Pedersen, M. 2017. Feasibility of energy justice: Exploring national and local efforts for energy development in Nepal. *Energy Policy*, 105, 668–676.

Kanase-Patil, A. B., Saini, R. P. & Sharma, M. P. 2010. Integrated renewable energy systems for off grid rural electrification of remote area. *Renewable Energy*, 35, 1342–1349.

Lai, C. S., Jia, Y., Xu, Z., Lai, L. L., Li, X., Cao, J. & McCulloch, M. D. 2017. Levelized cost of electricity for photovoltaic/biogas power plant hybrid system with electrical energy storage degradation costs. *Energy Conversion and Management*, 153, 34–47.

Lam, N. L., Smith, K. R., Gauthier, A. & Bates, M. N. 2012. Kerosene: A review of household uses and their hazards in low and middle income countries. *Journal of Toxicology and Environmental Health Part B, Critical Reviews*, 15, 396–432.

Legros, G., Rijal, K. & Seyedi, B. 2011. Decentralized energy access and the millennium development goals: An analysis of the development benefits of micro-hydropower in rural Nepal. Rugby, UK: United Nations Development Program.

López-González, A., Domenech, B., Gómez-Hernández, D. & Ferrer-Martí, L. 2017. Renewable microgrid projects for autonomous small-scale electrification in Andean countries. *Renewable and Sustainable Energy Reviews*, 79, 1255–1265.

Mahapatra, S. & Dasappa, S. 2012. Rural electrification: Optimising the choice between decentralised renewable energy sources and grid extension. *Energy for Sustainable Development*, 16, 146–154.

Maxwell, A. 2016. Post-HydroAysen, Chile's rivers remain at risk. *Available:* www.nrdc.org/experts/amanda-maxwell/post-hidroaysen-chiles-rivers-remain-risk [Accessed 20 February 2018].

Nasir, M., Khan, H. A., Hussain, A., Mateen, L. & Zaffar, N. A. 2018. Solar PV-based scalable DC microgrid for rural electrification in developing regions. *IEEE Transactions on Sustainable Energy*, 9, 390–399.

O'Connor, J. 2017. *Battery showdown: Lead-acid vs. Lithium-Ion* [Online]. Medium. Available: https://medium.com/solar-microgrid/battery-showdown-lead-acid-vs-lithium-ion-1d37a1998287 [Accessed 20 March 2018].

Palit, D. & Chaurey, A. 2011. Off-grid rural electrification experiences from South Asia: Status and best practices. *Energy for Sustainable Development*, 15, 266–276.

Pennise, D. 2005. Biomass pollution basics. Berkeley, California: University of California, Berkeley.

Prinsloo, G., Mammoli, A. & Dobson, R. 2017. Customer domain supply and load coordination: A case for smart villages and transactive control in rural off-grid microgrids. *Energy*, 135, 430–441.

Ray, S., Ghosh, B., Bardhan, S. & Bhattacharyya, B. 2016. Studies on the impact of energy quality on human development index. *Renewable Energy*, 92, 117–126.

Siddaiah, R. & Saini, R. P. 2016. A review on planning, configurations, modeling and optimization techniques of hybrid renewable energy systems for off grid applications. *Renewable and Sustainable Energy Reviews*, 58, 376–396.

Siemons, R. V. 2001. Identifying a role for biomass gasification in rural electrification in developing countries: The economic perspective. *Biomass and Bioenergy*, 20, 271–285.

Singh, J., Panesar, B. S. & Sharma, S. K. 2008. Energy potential through agricultural biomass using geographical information system – A case study of Punjab. *Biomass and Bioenergy*, 32, 301–307.

Singh, S., Singh, M. & Kaushik, S. C. 2016. Feasibility study of an islanded microgrid in rural area consisting of PV, wind, biomass and battery energy storage system. *Energy Conversion and Management*, 128, 178–190.

Sovacool, B. K. & Dworkin, M. H. 2015. Energy justice: Conceptual insights and practical applications. *Applied Energy*, 142, 435–444.

United Nations 2017. *Sustainable development goals 7: Affordable and clean energy* [Online]. Available: www.un.org/sustainabledevelopment/energy/ [Accessed 27 February 2018].

Watson, S. 2018. *Gas vs. electric stoves: Which is really more efficient?* [Online]. Available: https://home.howstuffworks.com/gas-vs-electric-stoves2.htm [Accessed 20 July 2018].

WHO 2016. Burning opportunity: Clean household energy for health, sustainable development, and wellbeing of women and children. Geneva: World Health Organization.

Wiemann, M. 2013. *Small wind in a developing world* [Online]. Elsevier. Available: www.renewableenergyfocus.com/view/32333/small-wind-in-a-developing-world/ [Accessed 27 February 2018].

Williams, N. J., Jaramillo, P. & Taneja, J. 2018. An investment risk assessment of microgrid utilities for rural electrification using the stochastic techno-economic microgrid model: A case study in Rwanda. *Energy for Sustainable Development*, 42, 87–96.

Wirfs-Brock, J. 2015. *Lost in transmission: How much electricity disappears between a power plant and your plug?* [Online]. Inside Energy Available: http://insideenergy.org/2015/11/06/lost-in-transmission-how-much-electricity-disappears-between-a-power-plant-and-your-plug/ [Accessed 20 February 2018].

World Energy Council 2018. *Hydropower* [Online]. London, UK. Available: www.worldenergy.org/data/resources/resource/hydropower/ [Accessed 20 July 2018].

7 Waste

Introduction

While none of the Millennium or Sustainable Development Goals explicitly addresses the management of waste, improved waste treatment could positively affect many different aspects of rural life. Currently, in many developing rural communities waste is either left untreated in piles sometimes on dry land and sometimes in waterways, or treated through open burning (Mihai & Ingrao, 2018). Both of these have the potential to cause significant environmental damage and pose a hazard to human health. Conversely, if treated appropriately "waste" has the potential to become a valuable resource contributing to income generation and offsetting the use of virgin resources to create products and generate energy.

Determining the most appropriate way to collect and treat waste depends on its composition. The World Bank reports that the composition of a country's solid waste is directly related to its GDP as shown in Table 7.1. In all but the highest income countries, organics make up the majority of the waste stream. In rural areas this increases to as much as 80 per cent of the total waste stream (Mohee, 2007).

Waste disposal options are generally presented in order of preference following a principle known as the waste hierarchy as shown in Figure 7.1. The hierarchy appears in many different forms from all over the world, but they are all basically the same. The version presented here is based on the European Union 2008 directive (Parliament and Council of the European Union, 2008).

Table 7.1 Current estimates of waste composition by income level

Income level	*Organic (%)*	*Paper (%)*	*Plastic (%)*	*Glass (%)*	*Metal (%)*	*Other (%)*
Low income	64	5	8	3	3	17
Lower middle income	59	9	12	3	2	15
Upper middle income	54	14	11	5	3	13
High income	28	31	11	7	6	17

Reference: Hoornweg and Bhada-Tata, 2012

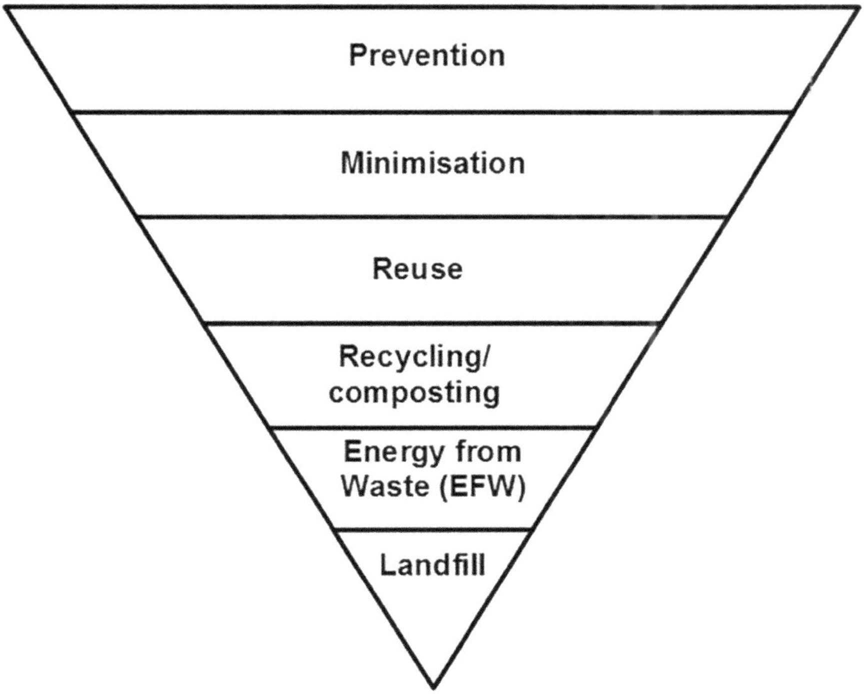

Figure 7.1 Hierarchy of waste disposal processes

- Prevention – covers various options such as replacing disposable products with reusable ones for example using plastic crates to transport goods instead of cardboard boxes. It can also refer to using an alternative product for example reading news online instead of buying a newspaper.
- Minimisation –reducing the quantity of material used in a product for example reducing the thickness of packaging to the minimum required for preventing damage to its product.
- Reuse – adapting a product for another purpose when it reaches the end of its useful life.
- Recycling/composting – processing waste to turn it into a new product (see Figure 7.2).
- EFW (incineration/ AD) – turning waste into energy through either incineration or anaerobic digestion.
- Landfill – a highly engineered system, where a hole in the ground is lined with impermeable materials to prevent liquid from leaching out. The space is fitted with pipes that collect and discharge the liquid to a treatment area, to remove impurities before releasing it into the environment. Landfills should also contain a gas collection system for the methane generated by the biodegradation of

Figure 7.2 Indian women recycling garbage

organic waste. Depending on the volume of gas generated this can either be used to generate energy for nearby premises or flared to turn it into carbon dioxide before it is released into the environment.

The distribution of waste disposal methods currently in use is shown in Figure 7.3 and Figure 7.4. Figure 7.3 indicates the total mass of waste disposed by waste disposal type, shown separately by World Bank income region. This clearly shows the significantly higher waste generation by high income countries. Figure 7.4 shows percentage of waste sent to each disposal method, also separated by income region. Two categories appear in these figures that are not part of the hierarchy.

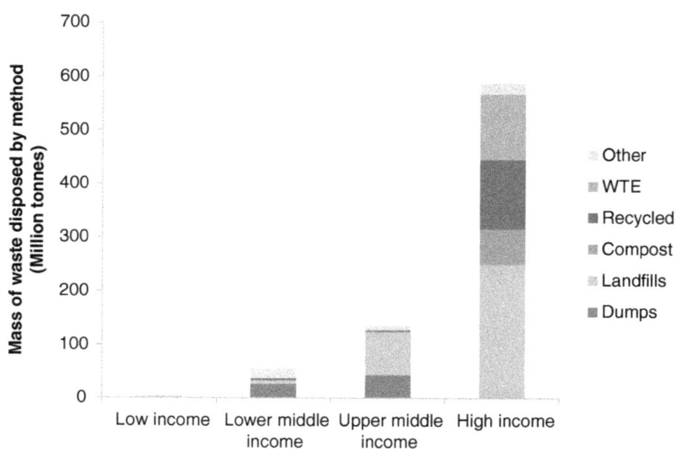

Figure 7.3 The total mass of waste by disposal method by countries' income range (adapted from Hoornweg & Bhada-Tata, 2012)

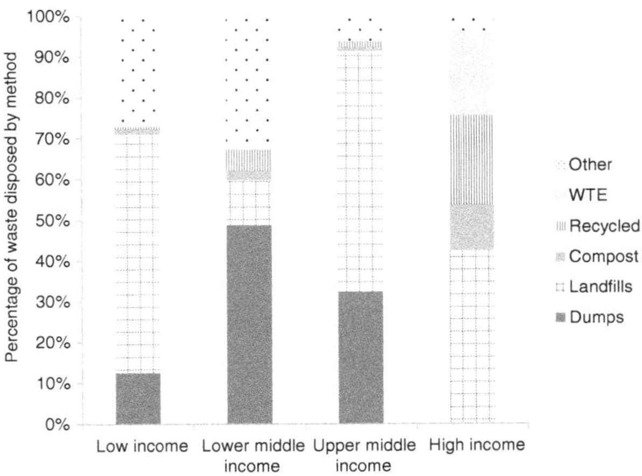

Figure 7.4 Percentage of waste by disposal method for different economic regions (adapted from Hoornweg & Bhada-Tata, 2012)

The category "other" is a result of different reporting methods from different regions, and unfortunately cannot be defined any more clearly. The category "dumps" refers to leaving waste in untreated piles. This practice poses a serious risk to public health, as well as an environmental hazard and is illegal in high income countries (although illegal dumps still do exist everywhere).

Figure 7.4 shows that in low- and middle-income countries landfilling and dumping are by far the main waste treatment methods in use, despite their rank as the worst possible option. The classification of landfill in some regions has also been brought into question, with suspicion that it would fall somewhere between a dump and a landfill meaning significant liquid and gas emissions could be entering the environment untreated. In rural areas, the high organic content of waste exacerbates the negative impacts of landfilling as it is their biodegradation that generates landfill gas. Landfill gas causes fires and explosions, odour nuisance, local air quality pollution, effects on vegetation and human health, and contains a significant proportion of greenhouse gas emissions. Although greenhouse gas emissions from organic materials are generally considered carbon neutral by the IPCC, this only applies if the emissions are in the form of carbon dioxide. When allowed to degrade in an anaerobic environment, the biodegradation process will produce methane, which has a radiative forcing potential 28–36 times higher than carbon dioxide (US EPA, 2017).

Unlike in urban areas, where municipal solid waste management is the single largest budgetary consideration of the local government, the low population density coupled with poor transport infrastructure in rural areas makes waste collection technically difficult and economically unfeasible. Thus, this largely organic waste stream is often left untreated creating an environmental burden and a health hazard (although not always as described in Case Study 7.1). Conversely, organic waste contains many nutrients that could potentially be beneficial to soils and has a huge energy potential. Thus, when treated through the most appropriate method it can go from being an environmental and health burden to a beneficial resource. In this chapter we will look at the two main options for treating organic waste: composting and anaerobic digestion and options for implementing these in a poor rural community.

Case Study 7.1 Mawlynnong: The cleanest village in India (Dutta, 2017)

The village of Mawlynnong, Meghalaya, India has been voted the cleanest village, not just in India but in Asia by various publications. Visitors to the village note several factors that distinguish it from the rest of India. Firstly there are bins, woven from bamboo, lining the streets preventing people from littering. Secondly, when locals notice litter on the street, they simply pick it up and put it in one of the bins without a second thought. Thirdly, villagers of all ages take turns sweeping the streets and public spaces. All of the organic waste collected is composted and used on soils. Each household in the village has beautifully kept front gardens. Other kinds of waste are reused wherever possible, for example, plastics are turned into planters or swings. The remaining waste is transported to an incinerator.

Although the village community did not set out to earn the title of cleanest village, they obviously take pride in their status. The fact that everyone in the community participates in maintaining cleanliness means everyone only has to spend a small amount of time on it, rather than a few people spending all day cleaning. Since being declared India's cleanest village Mawlynnong is now famous and provides a model for villages everywhere.

Composting

Organic wastes contain the nutrients required for plant growth, which are often depleted in soils where intensive monoculture farming is practised. (See Chapter 4 for more discussion on this.) Treating organic wastes through a composting process and returning it to the soil will improve soil quality. This in turn will help farmers to increase crop yields thereby increasing their income generation, while at the same time reducing the need to convert more land to agriculture. During the composting process the waste heats up to temperatures that kill pathogens making the resulting substance safe for handling and application to food crops.

There are various composting methods, some are very simple and can be practised at the household level. These include:

- Heap or pile – simply piling up organic waste is the simplest form of composting. It is only appropriate on a small scale to ensure air is getting to the waste, and it doesn't turn anaerobic. It is not recommended for areas with close neighbours as it may lead to odour nuisance. The lack of a physical barrier may also attract rodents or other pests to the compost heap, so awareness of these risks should be known.
- Digging a hole in the ground to pile the waste into is one step up from a simple pile.
- Creating an enclosure to house the composting pile can prevent some of the problems associated with heap/ pile compost such as limiting odour nuisance and limiting access for pests. The enclosure can be made of solid materials such as brick or wood, which can aid the increase in temperature in the compost but must be left partially open to ensure adequate air gets into the compost to prevent it from becoming anaerobic. A wire or mesh enclosure has the advantage of allowing plenty of air to get to the compost pile but will not aid the reaching of higher temperatures.
- A barrel or drum is a good option. These are usually mounted above the ground to improve aeration and limit access to pests. The barrel or drum also prevents odour nuisance. However, the compost must be turned periodically using a shovel to encourage aeration.
- A rotating barrel or drum is the best option because it has all the advantages of the barrel/drum system, without the need to turn the waste. Simply spinning the barrel mixes the waste, creating aeration with little physical exertion and without getting dirty.

Larger scale composting can be advantageous for larger farms or where individual households are unwilling or unable to manage their own compost. It can also provide an income generating opportunity through charging a waste collection fee at the start of the process, then selling the fertiliser material at the end of the process (see Chapter 2 for more discussion on this).

Larger composting systems can be:

- Passive pile – similar to small pile compost, the waste is piled up and left to mature with no work required. The compost takes a long time to reach maturity in this process, up to two years, and the types of waste that can safely be used in this method are limited.
- Windrow – where the waste is stacked in piles that are regularly turned to ensure aeration. The piles can be outdoors however, consideration should be given to local climate conditions. In rainy areas, shaping the top of the pile in a triangular way should allow water to run off, preventing the piles from excessive moisture, while in drier climates shaping the piles in a concave way will allow penetration of moisture preventing them from becoming too dry. The process is much faster than passive piles, usually five to ten weeks to reach maturity.
- Passive windrows – the waste is stacked in a similar way to windrows, only a bottom layer of a porous material such as straw is placed first, then fitted

with pipes. This creates aeration in the system, removing the requirement to physically turn the compost. The top of the pile is also covered with either finished compost or peat moss to minimise odour nuisance and prevent flies. There are limitations to the types of waste that can be treated by this method. The compost typically takes six to ten weeks to reach maturity.

- Aerated static pile – the structure of the waste is similar to passive windrows, only the network of pipes incorporated into the bottom layer are connected to fans. The fans are used both to suck air out of and blow air into the pile to provide greater control over aeration, temperature, odour etc. This is the fastest process of those listed here taking three to five weeks to reach maturity. The drawback is the process is much more complex and difficult to manage. Determining the aeration rates, size and number of fans, length diameter and types of pipes.

In most regions the decision whether to support small-scale on-farm composting or establish a large central composting plant is obvious. If composting is already widely practised or there is poor transport infrastructure small-scale composting will be the logical choice. Where there is good transport infrastructure and the majority of organic wastes are being dumped then establishing a large central composting plant may be the more appropriate option, although a survey of residents to determine their likely participation rates as both compost producers and users in either scheme should also be used to inform the decision. Han et al. (2015) surveyed households in a rural area of the Tibetan Plateau and found households somewhat reluctant to manage compost heap, yet 84 per cent of respondents expressed willingness to use waste derived compost on their crops. In Assam, India a small-scale composting initiative has been very successful as described in Case Study 7.2.

Case Study 7.2 Rotating compost bins from Assam Engineering College (Personal Communication, Pradip Baishya)

In the state of Assam, India a composting project spearheaded by Pradip Baishya from the Assam Engineering College has been very successful at diverting waste from traditional disposal methods. In Assam organic matter accounts for an estimated 70 per cent of the waste stream so it is a huge reduction in overall waste production.

The composting technique mixes dry leaves with the organic fraction of household waste in a rotary drum. The drum has holes drilled into it for aeration. Waste is loaded into the drum for ten days, and then rotated for the next 21 days. Next, the compost is removed from the drum and left to cure in the open air for seven days. Finally, it is sieved and is ready to be applied to soils.

The drums have been distributed all over the state of Assam both for single household use, and at larger scale composting sites such as educational institutions, vegetable markets and residential apartments blocks where waste generation exceeds 100 kg.

Follow-up surveys have been conducted with the composting drum users and the response has been very positive. Participants found managing the compost very easy and have developed it into a habit. In the city of Guwahati participants mentioned they would like to see the composting systems installed all over the city to help reduce the ever-growing waste management problem in the region.

Small-scale composting often raises concerns about whether the process is being managed properly. If the heaps are inadequately aerated there could be hundreds of mini-dumps generating copious amounts of methane. Concerns regarding the levels of contamination in the waste are also raised. Survey analyses into the management and composition of household level composting systems have been conducted in response to these concerns. Smith and Jasim (2009) surveyed compost management of household composting in west London and found they were all adequately managed in terms of minimising methane generation. Vázquez and Soto (2017) surveyed the composition of the output from 90 home composting bins in eight regions of Galicia, Spain and found that for the most part they were well managed. Heavy metal contamination was found in only 6 per cent of the systems, which was mainly attributable to households that put ash in their compost. At another household, grass cuttings that were contaminated with paint had been incorporated into the compost. Therefore, a basic campaign to raise awareness of appropriate wastes to include in compost could reduce heavy metal contamination risk. A comparison compost produced from a large industrial site to small-scale on-farm compost was conducted in Italy and it was found that the small on farm compost had less contamination than the industrial compost (Pergola et al., In Press). The authors suggest that the fact that the compost producer and user is the same person in small-scale composting generally leads to a better-quality compost.

Mohee (2007) reviewed composting projects that were undertaken in five different African countries. The projects included a mix of small-scale on farm composting, and larger composting systems. There were variations in the knowledge of participants at the start of the process. For example, in Zimbabwe many farmers were already informally composting some of their organic waste, while also using food scraps for animal feed etc. The project in this case aimed to educate farmers on how to optimise the composting process by monitoring the temperature and moisture content. It also educated the farmers in the best way to apply the compost to their crops.

In South Africa, Kenya and Mauritius, on the other hand, prior to the project composting was not widely practised, and the process was not well understood.

Thus, the project aimed to give farmers a deeper understanding of how nutrients in organic waste could improve their soil quality. In all regions the communities embraced the composting methods and developed them well. Long-term monitoring of the projects found that farmers continued to see increases in their crop yields over control plots that were not treated with compost, for several years after the composting scheme was implemented. The level of yield increase ranged from 15 to 30 per cent depending on the crop type.

Anaerobic digestion

Anaerobic digestion (AD), similar to composting, treats organic waste to produce a fertiliser. However, during biodegradation the waste is deprived of oxygen, producing a gaseous fuel with similar properties to natural gas. The process reduces the volume, mass and toxicity of organic wastes including manures (Roopnarain & Adeleke, 2017). In theory, AD is the ultimate sustainable system because it solves multiple sustainable development issues including:

1) It can be used to treat many different types of organic wastes including human faeces, manures and food scraps, preventing them from being released into the environment in an uncontrolled way. If they are released into the environment untreated they can:
    a Lead to environmental damage for example eutrophication of water ways, emissions of greenhouse gases and ozone depleting substances.
    b Create a health hazard. Organic wastes often contain pathogens that are toxic to humans, especially those that cause diarrhoeal diseases.
2) It produces a clean burning fuel that can be used for cooking and heating which:
    a Prevents the use of raw biomass burning for this purpose, which leads to poor indoor air quality. As described in Chapter 6 this can lead to many health problems especially lung disease. It also disproportionately impacts women who are more likely to perform indoor tasks such as cooking.
    b Prevents the use of forest products for biomass burning. Fuel wood can account for up to 54 per cent of deforestation in developing countries (Surendra et al., 2014).
    c Saves the time usually spent gathering biomass which has been reported to take up to six hours per day (Surendra et al., 2014). This task normally falls to women and children, depriving them of the opportunity to spend their time in education or an income generating activity.
3) The fuel could also be used with a generator to produce electricity, increasing access for remote communities who are not connected to their national grid, or have only intermittent electricity provision (Roopnarain & Adeleke, 2017).

4) The residual liquid and solids from the AD process are rich in nutrients such as phosphorus and nitrogen which can:

   a Offset the use of synthetic and/or mineral fertilisers where these are currently in use
   b Where no fertiliser is currently used it can improve soil quality, increasing crop yields and thus food security, while also potentially leading to increased income.

5) The use of small-scale AD systems means they can be operated on-farm, preventing the need to transport waste where there is poor transport infrastructure.

Two types of digester have been developed for the small-scale household AD. The fixed dome design, which was developed in China, and the floating drum digester, which was developed in India (Gunnerson & Stuckey, 1986). Both of these systems are made from a mix of materials such as bricks, steel and cement. These materials can be relatively expensive in many developing regions making the technology financially unattainable. A cheaper alternative is a polyethylene tube digester (Bond & Templeton, 2011). This system uses a cheap plastic material for the digestion chamber, making it more affordable. The downside of this material is that it is less durable and thus, has a shorter lifetime.

AD systems can operate as a batch, where the system is filled with waste, processed to completion, then removed completely (Rowse, 2011). After removal the process can then be repeated, in some instances retaining a portion of the residue in the system before the subsequent batch to aid the growth of bacteria can help speed up the process. The main advantage of this system is that if a contaminated batch enters the system, the contaminant will be removed after processing making the next batch uncontaminated. Systems can also operate continuously, where waste is fed into one end of the system and processed digestate is removed at the other end. Within the chamber the incoming waste will inevitably be mixed with the partially digested waste. The disadvantages of continuous systems are that if the system becomes contaminated the contaminant will be difficult to remove. In a batch process that contaminant will be removed with each batch.

So why isn't the use of AD more widespread?

High upfront costs

The main hindrance to widespread uptake of AD technology in developing regions is the high upfront cost. In fact, Bond and Templeton (2011) assessed the current state of biogas digester technology and found nearly all initiatives had required either government subsidies or funding from an NGO. Both China and India used government subsidies to disseminate AD technology, and it is estimated that there are currently 27 million systems in China and 4 million in India. However, not all governments have resources available to fund such a scheme. In these cases, obtaining funding under the clean development mechanism (CDM) could be a better solution. Finally, offering microfinancing loans to cover the start-up costs could be an option for some people, if the household is generating sufficient income to cover the loan repayments (Surendra et al., 2014). Roopnarain and Adeleke (2017) further suggest that to keep costs to a minimum, AD systems should be developed using cheap and locally available materials. They also suggest systems should be prefabricated. The downside of prefabricated systems is the one-size-fits-all approach could be difficult as farms produce widely different amounts of waste depending on their stocks.

Maintenance

While government subsidies and NGOs have seen millions of AD systems set up worldwide, follow-up assessment of these schemes has found a very high failure rate. Bond and Templeton (2011) reviewed schemes from all over the world and found that in many locations more than half of the AD systems were no longer operational after a relatively short time. Investigation into the reason for this noted that long-term success was often accompanied by long-term funding to provide system maintenance. AD systems are sensitive to:

- Moisture, often they require the addition of water which can be difficult in drier regions at certain times of the year.
- pH level, which sometimes requires the addition of an alkali such as calcium carbonate.
- Temperature which can be difficult to maintain in many regions that have seasonal weather conditions.
- Unwanted compounds in the waste stream, these can be difficult to avoid, and can lead to the formation of corrosive gases, which can damage the equipment of the AD system, especially the generator if it is used for electricity production.

To overcome system failures the first step is to ensure that the system is as simple as possible to operate. Secondly, installation must be accompanied with at least some basic training for the household members. Accompanying this with more detailed training for a few members of the community to provide the region with higher level expertise would also be advised. This would have the added bonus of creating a job for the members of the community selected for the course. The difficulty may be whether the owners of the AD systems can afford to pay the maintenance costs, or whether financing will be made available from other sources such as government subsidies.

Even when maintenance expertise is available, any repairs to the digester systems may be hindered by lack of access to the materials required to perform the repairs. Developing a digester made from cheap and locally available materials is therefore essential.

The necessity to store the gas that is produced can also be difficult as it is highly explosive. The gas is essentially odourless, so it can be impossible to detect if it is leaking. This could lead to explosions from a simple ignition source. Exposure to high concentrations of the gas can also lead to asphyxiation.

Availability of feedstocks

The sensitivity of AD systems extends to the quantity and composition of feedstock entering the system. Sizing a digester to ensure it is appropriate for the quantity of organic waste that it will be treating is good practice, although this could make the installation complex and overly expensive. It also removes the option of reducing installation costs by establishing a prefab system. Using census survey data to characterise the organic waste generation of a typical household in the region would be a better way to determine appropriate sizing. This could then allow for prefabrication of an appropriate system. Another issue found in previous research was that many farmers believed that the system operates solely on cow manure (Roopnarain & Adeleke, 2017). In reality, the system performs optimally when the feedstock contains a mix of organic waste types. Raising awareness of this will be important for long-term operational success.

The emphasis on the use of manure raises another issue in some regions where livestock are free to roam during the day and are only penned overnight. This means their manure is all over the place and collection can become laborious and time-consuming. Similar to composting, an AD system could be operated at a community scale to help overcome the problem of lack of feedstock (Roopnarain & Adeleke, 2017). Previous attempts at implementing a community scale AD system have resulted in difficulties with the users taking responsibility for the system. For this type of system to be successful communities must work together and share responsibilities.

Awareness of the technology

Another hurdle to the widespread uptake is lack of awareness of the technology. There are two components to this: firstly, there are many people who have simply never heard of it; secondly, people are aware that it has been tried and failed in many locations and so are wary of investing in the technology. Raising awareness of the technology and promoting any advances that have overcome the system failures that plagued the AD initiatives of the past will be necessary to generate enthusiasm to invest in AD technology.

Recommendations for best practice

The use of on-farm composting should be encouraged for all smallholder farmers. Subsidies for a composting drum or barrel and training materials on how to manage and use compost would be essential to maximise the benefits of this initiative.

Where large scale composting is more appropriate assessment of the local area and availability of expertise should be taken into consideration. The process should be kept as simple to manage as possible. Training into maintaining aeration and ensuring temperatures reach appropriate levels to destroy pathogens will be an essential component of ensuring the safe operation of the system and use of the resulting composting material.

While theoretically anaerobic digestion would be a superior option to composting for treating organic waste, due to the generation of a fuel for energy, the complexity of maintaining the system means there is a high risk of failure. Therefore, we would generally recommend implementing a composting system unless there are local experts who can aid maintenance problems associated with AD to ensure the long-term success of these systems.

References

Bond, T. & Templeton, M. R. 2011. History and future of domestic biogas plants in the developing world. *Energy for Sustainable Development*, 15, 347–354.

Dutta, S. 2017. *This is Asia's "Cleanest Village" and it is in India*. Banega Swachh India. Available: https://swachhindia.ndtv.com/asias-cleanest-village-india-5039/ [Accessed 23 July 2018].

Gunnerson, C. C. & Stuckey, D. C. 1986. Anaerobic digestion: Principles and practices for biogas systems. *Integrated Resource Recovery*. Washington DC: The World Bank.

Han, Z., Dan, Z., Shi, G., Shen, L., Xu, W. & Xie, Y. 2015. Characteristics and management of domestic waste in a rural area of the Tibetan Plateau. *Journal of the Air and Waste Management*, 65, 1365–1375.

Hoornweg, D. & Bhada-Tata, P. 2012. What a waste: A global review of solid waste management. *In*: Bank, T. W. (ed.), *Urban development knowledge papers*. Washington DC: The World Bank.

Mihai, F.-C. & Ingrao, C. 2018. Assessment of biowaste losses through unsound waste management practices in rural areas and the role of home composting. *Journal of Cleaner Production*, 172, 1631–1638.

Mohee, R. 2007. Waste management opportunities for rural communities. *In*: FAO (ed.), *Agricultural and Food Engineering Working Document*. Rome: Food and Agricultural Organization of the United Nations.

Parliament and Council of the European Union 2008. Directive 2008/98/EC on waste. *Official Journal* L 312, 3–30.

Pergola E. M., Persiani, A., Palese, A. M., Meo, V. D., Pastore, V., D'Adamo, C. & Celano, G. In Press. Composting: The way for a sustainable agriculture. *Applied Soil Ecology*.

Roopnarain, A. & Adeleke, R. 2017. Current status, hurdles and future prospects of biogas digestion technology in Africa. *Renewable and Sustainable Energy Reviews*, 67, 1162–1179.

Rowse, L. E. 2011. *Design of small scale anaerobic digesters for application in rural developing countries*. Master of Science, University of South Florida.

Smith, S. R. & Jasim, S. 2009. Small-scale home composting of biodegradable household waste: Overview of key results from a 3-year research programme in West London. *Waste Management and Research*, 27, 941–950.

Surendra, K. C., Takara, D., Hashimoto, A. G. & Khanal, S. K. 2014. Biogas as a sustainable energy source for developing countries: Opportunities and challenges. *Renewable and Sustainable Energy Reviews*, 31, 846–859.

US EPA 2017. *Understanding global warming potentials* [Online]. Washington DC. Available: www.epa.gov/ghgemissions/understanding-global-warming-potentials [Accessed 3 May 2018].

Vázquez, M. A. & Soto, M. 2017. The efficiency of home composting programmes and compost quality. *Waste Management*, 64, 39–50.

8 Water and sanitation

Introduction

The human body is approximately 60 per cent water. It is used in digestion, absorption, circulation, creation of saliva, transportation of nutrients and maintenance of body temperature (US Geological Survey, 2016). Humans need to drink water every day so that all of these bodily functions can perform. Yet 663 million people do not have access to clean drinking water (United Nations, 2018) and at least 1.8 billion people use a source of drinking water that is faecally contaminated. Human waste contains many common bacterial infections and can spread other infectious diseases, leading to illness or in some cases deaths. The WHO estimates around 1000 children die every day from drinking contaminated water (United Nations, 2018). Faecal contamination of water can be prevented by sanitation facilities and wastewater treatment. Without sanitation facilities, human waste will also end up in soils, where it can accumulate in foods, and diseases contained in human waste can be spread by insects (Centers for Disease Control and Prevention, 2015).

Aside from drinking, humans need water for many other purposes such as cleaning, cooking, irrigation and energy generation. It is essential to our survival in many ways, yet, one fifth of the global population lives in areas with physical water scarcity and even more live in areas affected by economic water scarcity (Dell'Angelo et al., 2018). Water resources can be considered "blue" which includes all forms of freshwater storage such as lakes, rivers, aquifers or "green" which accounts for water stored in soils. Irrigation represents the largest human use of blue water with approximately 69 per cent of annual withdrawals and approximately 92 per cent of annual consumptive blue water use. In some areas, large tracts of land are being appropriated by international corporations for crop production. While the loss of land to the local residents may be justifiable, the use of blue water resources by these corporations can set a limit on the ability of the residents to improve their quality of life. It is often argued that the locals are currently practising a rain-fed crop production system but diverting water resources removes the opportunity for them to upgrade their agricultural practises in the future. A practice that is becoming more essential as periods of drought are becoming longer and more frequent.

Figure 8.1 Collecting and transporting water in India

It is not just humans that require water for survival, nearly all species of plants and animals require some level of water input to thrive. Livestock require clean water to thrive and produce good quality yields of milk or meat depending on their purpose. The precise amount depends on their stage of growth cycle, their size and whether they are lactating, as well as the ambient temperature and humidity of their location. The quality of the water will impact their consumption, for example bad-tasting or smelling water may put animals off drinking, which will impact their ability to produce milk, or impact their quality of their meat etc. Feed with high moisture content will reduce the amount of water animals require for drinking.

Ward (2015) lists the typical water consumption requirements of common livestock species, at different stages of their growth cycles.

In this chapter we begin with a discussion on bringing access to potable water to rural communities, followed by sanitation facility options. Then we discuss wastewater treatment methods. Finally, we discuss options for managing water resources in a more sustainable way ensuring all water services are provided, not just for today but also for future generations.

Provision of potable water

The provision of water will depend on available sources. Table 8.1 shows the main components of a water supply system. Note, not all of the components will be necessary for all systems and will depend on location specific factors.

In rural areas many people are required to travel to collect water and carry it back to where it is needed as illustrated in Figure 8.1. Piped water would significantly improve a person's quality of life, saving them from frequently performing this laborious task. However, with small populations the provision of a piped drinking water supply may not be economically feasible (Guo & Englehardt, 2015).

Table 8.1 Components of a water supply system

Component	Description
Spring intake structures	Underground structures that collect water issuing from natural springs and feed it into a pipeline. The structure should prevent contamination, ensure hygienic quality of the raw water, and remove suspended particles. It should also contain an overflow drain for when yield exceeds demand.
Bored deep well	An effective, safe and economical way of collecting groundwater.
Dug well	A circular, vertical, concrete structure for accessing groundwater. Simply constructed by digging a pit.
Pumping stations	These deliver water to tanks or distribution systems where there is either no gravity head, or it is insufficient to provide the necessary flow.
Aeration station	These are required where raw water (usually groundwater) has low dissolved O_2, or contains high concentrations of metals such as Mn or Fe, or has a high concentration of CO_2. A chemist should be consulted to determine the best method of metal removal.
Sedimentation tank	To remove suspended particles from raw water.
Water main ancillary chambers	Ancillary structures on water pipelines such as a pressure head breaking chamber, air and drain valve chambers and a measuring chamber.
Sand filtration	Water is passed through sand to remove contaminants, such as suspended particles. Mostly used for surface water sources, but can be necessary for groundwater depending on the quality.
Disinfection	This step removes bacteria from the water, preventing disease and illness. Exposing the water to sodium hypochlorite for 30 minutes is the usual method. The location of the disinfection stage can vary depending on the location of the various elements of the water distribution system and should be decided on a case by case basis.
Water tanks	These tanks store water, so that it is available when demand exceeds flow. It can also be used for emergencies, for example to fight fires. The amount of excess should be sufficient to prevent an interruption to the water supply in the event of the water source being temporarily disconnected from the distribution system.
Water-tank valve structures	To control and regulate water flow to and from storage tanks.
Other network structures	Fire hydrants, standpipes, house connections etc.

Reference: WHO, 1986

In some areas, alterations to the landscape have caused above ground water sources to dry up. However, it is often possible to reverse this situation as described in Case Study 8.1. Other options to bring water closer to villagers include accessing groundwater, through wells, or rainwater harvesting (Wang et al., 2014). While both of these methods are possible, it should be noted that the quality of this water is questionable and must be treated before drinking. Options for treating water to bring it up to drinking water quality include boiling, chlorination, dilute hypochlorite solution, porous ceramic filtration and combined flocculant-disinfectant powder. However, these options may not be feasible in many regions. The energy required to boil water could be beyond the means of many rural households. The cost of water purification tablets and other disinfectants may be prohibitively expensive.

Case Study 8.1 Selah Ranch, Texas (Masters, 2017)

In the early 1970s J. David Bamberger purchased this 5500 acre plot of land in Texas. At that time the area had been overused and was completely dry. David had seven water wells dug to try and find some water, up to 500 feet deep and found nothing.

The land had been used for grazing, removing all of the grasses that were natural to the landscape. Instead of replacing these, the land was covered with trees such as cedar. Bamberger had many of the trees removed and replanted the grasses that were native to the region.

Two and a half years later the first spring appeared. The reason, previously when there was rainfall, the soil structure was too dense to absorb the water, and therefore it would run off. The root system of the grasses allowed water to percolate through the soil and reach the aquifers. These slowly began to fill up, eventually overflowing leading to the replenishment of the springs. As time went on, wildlife returned to the area, and it is now a biodiverse nature reserve.

A more affordable option could be solar disinfection, where the water is exposed to solar radiation. The effectiveness of the treatment was tested by Dessie et al. (2014) in Ethiopia. They found that the treatment was effective when the water's depth was shallower, the turbidity of the water was lower, the container was a black plastic bottle, and the intensity of the solar radiation was higher (i.e. it was less effective on cloudier days). Other factors such as bottle placement, exposure and pH did not show any relationship to microbial inactivation. When the conditions were right, microbes were effectively inactivated and did not regrow. The exposure time was less than four hours. Thus, this could be a practical option for regions with higher solar radiation levels, such as those closer to the equator. However, the experiment indicates that the conditions will be site specific and so experiments must be done in the region to ensure the conditions are optimal and effective inactivation of the microbes occurs.

Sanitation facilities

Coming into contact with human waste is a major cause of illness globally, which could easily be prevented by access to hygienic sanitation facilities. The provision of sanitation facilities can also improve educational opportunities for girls, who may find it difficult to continue attending a school with no sanitation facilities after they begin menstruating. Open defecation is still practised in many regions. While in remote areas with very small populations the environment may be able to tolerate some level of human waste, it is still not recommended as it will be difficult to avoid coming into contact with the waste, especially where there is a lack of understanding of the health hazard it poses. Wang et al. (2014) noted that in some parts of Africa people defecate into plastic bags and then dump these at the side of the road at night. Other regions technically have sanitation facilities; however, these cannot always be considered hygienic because they do not prevent people from coming into contact with the waste. A hanging latrine, for example, is where a platform is built directly over a waterway with a hole allowing defecation directly into the water. Clearly, this will lead to water contamination and is especially concerning where it is practised around a waterway that is also used for drinking water, fishing, or other purposes that will bring it into contact with humans (Stauffer, 2011). Similarly, a bucket latrine, where a bucket is used in the home and then manually emptied when it is full, either by a member of the household or the council cannot be considered hygienic. Where the council collects the waste, this is sometimes done sporadically meaning the buckets will be overflowing by collection time (WASHplus, 2010).

Both the MDG and SDG contain a goal relating to sanitation. The achievement of their sanitation goal is based on whether the facilities are considered improved or unimproved. Improved sanitation facilities are defined as those which ensure separation of human waste from human contact. Another organisation devoted to sanitation

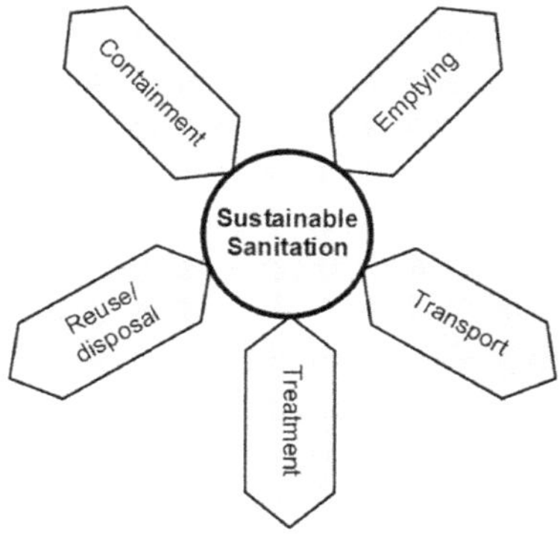

Figure 8.2 Sustainable sanitation value chain

provision is The Sustainable Sanitation Alliance. They define sustainable sanitation as a system that complies with their five-step value chain shown in Figure 8.2.

There are many different types of sanitation facilities that fit the definition of "improved":

Toilets that flush into piped sewers are the most common type of sanitation in many developed countries. This can be a sustainable option if the water is piped to a treatment facility and has any contaminants removed before release into the environment. An even better option is treating the water to a high standard, so that it can be reused, rather than being discharged into rivers or oceans. The sludge that remains after treatment can be anaerobically digested for energy generation, and then applied to soils as a complement to fertilisers (although the quality is not as high as other compost options, and in some cases, it is more valuable for its water content in dry seasons, and for improving soil structure to increase is water holding capacity and prevent erosion, rather than its nutrient content). However, there are regions where sewage is discharged into rivers or oceans untreated which can lead to many environmental hazards, disease outbreaks and loss of income for fishing and tourism industries.

In an area with a smaller population or terrain that makes digging sewage pipes unfeasible, flushing to septic tanks or pit latrines may be a more practical option. In these, the water is allowed to filter through the soil, which theoretically will naturally remove contamination before the water reaches aquifers. Again, disposal options for the remaining sludge will be a key part of determining the sustainability of the overall system.

However, in developing areas, the use of water to flush toilets can be adding more stress to already stretched water resources and may not be a feasible option. Other non-water using systems can be just as hygienic, which include ventilated improved pit latrine, pit latrine with slab and composting toilets.

Pit latrines are where a deep hole is dug in the ground, which may need to have the sides reinforced to prevent collapse (Mara, 1985). This simplest form of a latrine is the worst option and considered "not improved" under the MDG/SDG guidelines. Adding a platform or slab at the top of the latrine, that covers over most of the hole preventing the risk of falling in, and controlling some of the odour nuisance is considered an improved design. There are two broad options for latrines, they can be single or double pit. Once a single pit latrine is full, it will be closed off and a new facility will need to be dug out. With a double system, after the first hole is filled, the second hole can be used, allowing the waste in the first hole to degrade, naturally killing pathogens. After a suitable amount of time, the first latrine can be dug out and the remaining waste, which will now be clean and nutrient-rich compost, can potentially be returned to the soil for fertilisation or disposed of in a landfill. The first pit can then be used again once the second latrine is full and so the process continues, avoiding the necessity of having to dig further pits. It is important to avoid groundwater level when digging the latrine. If the groundwater table is high, then it may be necessary to dig a shallower pit, and then raise the platform above ground level.

Composting toilets operate similarly to a pit latrine, except instead of just digging a hole, a tank is used to capture and store the waste (Personal Communication, David O'Brien). This can be a more appropriate option in regions with higher rainfall, a high groundwater table, or coastal areas. Pit latrines are exposed to rain and in periods of heavy rainfall could flood, spreading the waste throughout the surrounding area. This would be a very unhygienic situation and must be avoided. In a composting toilet, the tank is designed to prevent the inflow of water, avoiding this situation.

Providing people with hygienic sanitation facilities may not be enough. In the past, governments and charitable organisations have built hygienic sanitation facilities for a village. On a follow-up visit they found the facilities were unused, and people were still practising open defecation (Royte, 2017). Therefore, especially in areas that are still practising open defecation, providing the community with education on not just how to use the facilities, but the associated health risks of not using them is essential. Community members will also need training in the maintenance of the sanitation system especially where they have not used it before. Case Study 8.2 shows an example of how community engagement empowered a rural community in Papua New Guinea to manage their own sanitation facilities.

Case Study 8.2 Composting toilet in Sipaia, Papua New Guinea (Personal Communication, David O'Brien)

In 2012 a travelling studio from the University of Melbourne teamed up with PNG University of Technology, Lae to build three innovative composting toilets in the village of Sipaia, PNG.

Previously, the residents were using pit latrines, which often flooded, allowing faecal waste to flow through the village. The toilet operates on a similar principle to a double system pit latrine, only instead of digging holes into the ground to fill with waste, two tanks are installed. This avoids the risks of pits filling with water during rain or high tide, and preventing sewage waste from flowing through the village. The first tank is used until full, before moving onto the second tank. The decomposition process takes six months, which should be sufficient time for the second tank to be used before requiring the first tank again. After the six months the bulk of the waste has evaporated, and the remainder has biodegraded, destroying pathogens. This dry waste is disposed of on bushland, and the tank is reused. The new toilets are unaffected by floods and tides, and each is fitted with a solar powered light to make them safer to use at night. The roof of the toilet facility has a rainwater collection system providing water for handwashing – the toilet system does not require any water. The materials used for construction are locally available and inexpensive.

The construction of the toilets was a joint effort between the studio team and members of the community. Members of the community were then trained in the maintenance of the toilets and left to manage them.

In July 2013 members of the studio from University of Melbourne revisited the village to check on progress and used this opportunity to establish further research.

In 2016, another studio returned to Sipaia. They found the toilets were being used and fulfilling their needs. At this time the studio team constructed two more toilets together with members of the community (mainly members of the community).

They also took this opportunity to check over the toilets built in 2012 and replace some components that were showing signs of wear and tear with more durable materials.

During a follow-up visit in 2017, the University of Melbourne team ran a training programme for members of the local community on how to carry out the construction of the composting toilet, including things like budgeting and sourcing materials. Another toilet was built with no assistance from the University of Melbourne team showing the project had successfully given the local community control over their sanitation facilities. A truly sustainable result.

Treatment of wastewater

In areas where flushing sanitation facilities exist, it is essential that this water is treated prior to discharge into natural waterways. Water is also used in many commercial and industrial processes, including agriculture. During processing the water is often polluted with substances which can often be toxic to humans and other species. Yet, in many regions wastewater is discharged into rivers, oceans and other waterways untreated. It once seemed that the oceans and rivers were so huge they could absorb all the pollution and dissipate it to safe levels. However, we now know the impacts of pollutants can be far reaching. While the cost of treating wastewater seems expensive, the costs of not treating it can be even more prohibitive. For example, in Peru in 1992 there was an outbreak of cholera due to poor sanitation and inadequate disinfection of drinking water. Fishing and tourism accounted for 34 per cent of Peru's GDP at that time, and the loss of income combined with the health costs were estimated at 1 billion USD, ten times the annual national budget for water supply and sanitation (UNEP, 2004).

UNEP in cooperation with WHO, UN-Habitat and WSSCC devised a strategic action plan on municipal wastewater adopted at the UNEP/GPA intergovernmental review meeting in Montreal (UNEP, 2004). This action plan outlines ten key stages for local and national action on municipal wastewater:

1 Secure political commitment and domestic financial resources
2 Create an enabling environment at national and local levels
3 Do not restrict water supply and sanitation to taps and toilets
4 Develop integrated urban water supply and sanitation management systems also addressing environmental impacts

5 Adopt a long-term perspective, taking action step-by-step, starting now
6 Use well-defined timelines, and time-bound targets and indicators
7 Select appropriate technology for efficient and cost-effective use of water resources and consider ecological sanitation alternatives
8 Apply demand-driven approaches
9 Involve all stakeholders from the beginning and ensure transparency in management and decision-making processes
10 Ensure financial stability and sustainability.

Typical wastewater treatment facilities in developed countries involve a series of processes to remove impurities of the water as outlined in Figure 8.3. There are other optional steps that can be added to the process depending on the required quality of the water.

Establishing this type of wastewater treatment plant is expensive. The operation of these treatment plants is also expensive requiring pumping, aeration, chemicals and transportation. They are also highly technical and require skilled personnel for successful operation. Finding a suitable location for a plant can also be challenging. Therefore, in a developing rural community more passive wastewater treatment options may be more appropriate. These are known as natural wastewater treatment systems. The most widely used of these are constructed wetlands. Many species of flora have been identified which can effectively remove effluent from wastewater. For example, water hyacinth has been shown to effectively remove total suspended solids, biochemical oxygen demand, phosphorus and nitrogen (Laitinen et al., 2017). A LCC comparison between a technical wastewater treatment facility and a constructed wetland with water hyacinth found that the wetland was half the cost of the treatment plant, but equally as effective at removing contaminants from the water. The operation of the wetland system was also far less expensive and did not require technically skilled personnel. The main risk identified for a constructed wetland is the choice of plant species, as these can sometimes spread into the surrounding area and become a weed for nearby plantations if not carefully controlled.

The water output from a wetland system can sometimes contain phosphorus and nitrogen, making it tempting to divert it for irrigation. However, the constructed wetland cannot necessarily remove heavy metals, so it is not recommended that the outputs are used for irrigating food crops, but could be used for industrial crops (Ohandja et al., 2012). This would also prevent discharging excess nitrogen and phosphorus into waterways where they could potentially cause eutrophication.

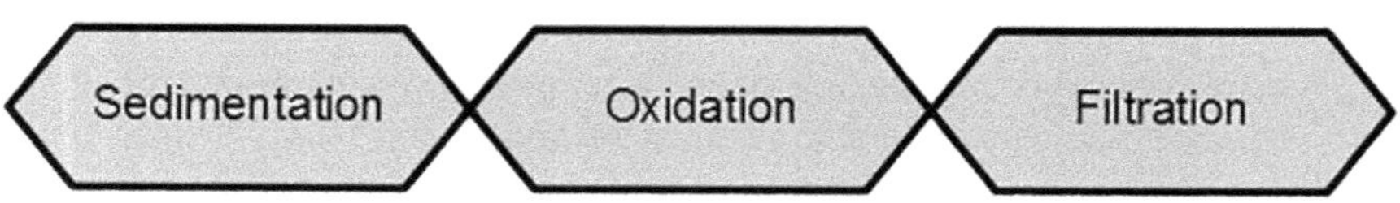

Figure 8.3 Main stages of a wastewater treatment process

Balancing water resources

The earth's precious freshwater resources are essential for many purposes, which are often competing for use from the same source. Irrigation is essential for farming to produce high yields that maximise production per unit area; hydropower is also the leading form of renewable energy, currently producing 16 per cent of the world's electricity, and will be an important component of a sustainable energy future, in fact, all forms of energy generation will require water use at some point in their life cycle; Fishing is the main source of income for many people, and can be very negatively impacted by other people's use of water systems, particularly for irrigation, as in the Aral Sea (see Chapter 4) where a fishing industry employing thousands was devastated by the poorly managed irrigation of the surrounding lands to grow cotton. WHO estimates that 1.7 billion people are currently living in river basins where water use exceeds recharge (United Nations, 2018).

All of these different water requirements are often managed by different organisations or government departments, competing against each other for the same water. All of these institutions have a basic shared purpose – to improve quality of life. They should be working together to ensure water resources are being used efficiently, and that they are not using more water than can be replenished, so that the water source does not dry up. Distefano and Kelly (2017) projected water availability up to 2100 based on the four RCPs of the IPCC and noted that for countries that are already experiencing water shortages this situation is only set to get worse unless we significantly increase the efficiency of our water use. The most influential factor contributing to water scarcity was not climate change, but the projected increases in wealth. Branche (2017) proposes water resources management should follow the SHARE concept:

- **S**ustainability approach for all users
- **H**igher efficiency, equity among sectors
- **A**daptability for all solutions
- **R**iver basin perspectives for all
- **E**ngaging stakeholders.

As an example, the Durance-Verdon River Basin in France has 17 dams and 30 hydroelectricity plants. The rivers also supply drinking water, irrigation, industry and tourist activities, as well as being home to an important ecosystem. The system was designed to supply all of these competing water uses from its conception in 1955. Although it is ultimately operated by the hydroelectric companies, they are obligated to supply all the other water requirements under the terms of their construction agreement (Branche, 2017). The dam has also helped to regulate water flows in the area through both droughts that have occurred in the Provence region, and flooding that occurred in the Durance River. More recently, as the impacts of climate change are starting to affect water flows, a study has investigated how the river system may be impacted, leading to further cooperative approaches to improve water use efficiency. One way they do this is to fund farmers in purchasing more efficient targeted irrigation equipment to minimise their water needs.

Recommendations for best practice

Access to clean drinking water is essential for survival and its provision to rural communities must be a priority. Ensuring water is free from faecal contamination through provision of sanitation facilities is an essential component of this.

It is recommended that sanitation facilities that do not use water are developed preferentially, especially in areas already struggling with water scarcity. The use of double pit latrines with a platform should be satisfactory except in regions that have high groundwater, are prone to flooding or are located near tidal areas. In this case the use of a composting toilet system is recommended.

In areas with no current wastewater treatment systems, implementation of a "natural" wastewater treatment system such as a constructed wetland is recommended, especially where the surrounding waterways are supporting a fishing industry that could be negatively impacted by the discharge wastewater.

The management of water resources needs to be a cooperative effort between all water users and the government, rather than a competitive situation. The management must ensure that annual water consumption does not exceed the water replenishment rate, so that freshwater resources will still be available for future generations.

References

Branche, E. 2017. The multipurpose water uses of hydropower reservoir: The SHARE concept. *Comptes Rendus Physique*, 18, 469–478.

Centers for Disease Control and Prevention 2015. *Toilets and latrines* [Online]. Atlanta, Georgia. Available: www.cdc.gov/healthywater/global/sanitation/toilets.html [Accessed 8 March 2018].

Dell'Angelo, J., Rulli, M. C. & D'Odorico, P. 2018. The global water grabbing syndrome. *Ecological Economics*, 143, 276–285.

Dessie, A., Alemayehu, E., Mekonen, S., Legesse, W., Kloos, H. & Ambelu, A. 2014. Solar disinfection: An approach for low-cost household water treatment technology in Southwestern Ethiopia. *Journal of Environmental Health Science and Engineering*, 12(1), 25.

Distefano, T. & Kelly, S. 2017. Are we in deep water? Water scarcity and its limits to economic growth. *Ecological Economics*, 142, 130–147.

Guo, T. & Englehardt, J. D. 2015. Principles for scaling of distributed direct potable water reuse systems: A modeling study. *Water Research*, 75, 146–163.

Laitinen, J., Moliis, K. & Surakka, M. 2017. Resource efficient wastewater treatment in a developing area – Climate change impacts and economic feasibility. *Ecological Engineering*, 103, 217–225.

Mara, D. D. 1985. Ventilated improved pit latrines: Guidelines for the selection of design options. *In*: Program, U. N. D. (ed.), *Technical advisory group discussion paper 4*. Washington DC: The World Bank.

Masters, B. (Director) (2017). *50 years ago, this was a wasteland. He changed everything*. YouTube: National Geographic. Available: www.youtube.com/watch?v=ZSPkcpGmflE [Accessed 13 September 2018].

Ohandja, D.-G., Donovan, S., Castle, P. & Voulvoulis, N. 2012. Regulatory systems and guidelines for the management of risk. *In*: Plant, J. A., Voulvoulis, N. & Ragnarsdottir, K. V. (eds.), *Pollutants, human health and the environment: A risk based approach*. London: Wiley-Blackwell.

Royte, E. 2017. Nearly a billion people still defecate outdoors. Here's why. *National Geographic*.

Stauffer, B. 2011. *Overhung latrine* [Online]. Willisau, Switzerland: Sustainable Sanitation and Water Management Toolbox. Available: www.sswm.info/water-nutrient-cycle/wastewater-treatment/hardwares/user-interface/overhung-latrine [Accessed 27 March 2018].

US Geological Survey 2016. *The water in you* [Online]. Washington DC. Available: https://water.usgs.gov/edu/propertyyou.html [Accessed 20 July 2018].

UNEP 2004. Guidelines on municipal wastewater management. The Hague, The Netherlands.

United Nations 2018. *Sustainable Development Goals 6: Clean water and sanitation* [Online]. United Nations. Available: www.un.org/sustainabledevelopment/water-and-sanitation/ [Accessed 15 March 2018].

Wang, H., Wang, T., Zhang, B., Li, F., Toure, B., Omasa, I. B., Chiramba, T., Abdel-Monem, M. & Pradhan, M. 2014. Water and wastewater treatment in Africa – current practices and challenges. *Clean – Soil, Air, Water*, 42, 1029–1035.

Ward, D. 2015. Water requirements of livestock. Ontario: Ontario Ministry of Agriculture.

WASHplus 2010. *Zimbabwe – Chiungwiza's "bucket system"* [Online]. Sanitation Updates. Available: https://sanitationupdates.wordpress.com/tag/bucket-latrines/ [Accessed 27 March 2018].

WHO 1986. Standard model designs for rural water supplies. Copenhagen: World Health Organization.

9 Transportation

Introduction

Without transport infrastructure rural communities can be essentially cut-off from the outside world, impeding their ability to pursue Sustainable Development Goals. Children cannot access schools to improve their educational attainment; farmers cannot access markets to sell their produce, suppressing the motivation to increase yields; employment opportunities are limited; access to essential services, such as healthcare facilities, are hindered. The village is also inaccessible to external services such as firefighters, ambulances and waste collection. Receiving goods in the village such as farming equipment or building materials can be severely limited often forcing residents to purchase goods locally at an elevated price. In regions with limited or no electricity travelling to visit family and friends may be the only way of staying in touch and finding out news (Johnston, 2007). Thus, the provision of transport infrastructure can open up multiple opportunities for rural communities to improve their quality of life.

For developing economies transport infrastructure development can be limited by budgetary constraints. Kilkenny (2010) argues that the expense of connecting sparsely populated rural communities is economically unjustified and questions the value of retaining small villages. Figure 9.1 for example, illustrates the vastness

Figure 9.1 Road construction at the village of Lamayuru, Leh Ladakh, India

of developing road infrastructure in the remote mountainous regions of Northern India. Conversely, Rönkkö et al. (2017) argue that rural communities provide resources not found in urban centres and have inherent value that cannot be discounted. The discussion in this book has hopefully shown that there are ample opportunities for rural villages to flourish, however, to do so they will require transport links. The first section of this chapter will discuss strategies for deciding on the most beneficial way to develop infrastructure within budget.

Without regular maintenance roads soon fall into disrepair, often becoming impassable. This will render the rural community once again isolated from the outside world and the initial investment will have been wasted. Yet in many developing regions maintenance is sorely lacking. The second section of this chapter will explore the reasons for this and describe examples of how these have been overcome.

While building and maintaining roads for motorised vehicles is an important component of transport infrastructure, it needs to cater to all modes of transport using the road. The third section of this chapter will explore the distribution of transport modes in developing economies and the dangers of ignoring the requirements of non-motorised transport. It will then look at ways these can be incorporated into road design. Many rural communities rely on public transport such as buses, especially for longer journeys. The final section of this chapter will discuss the provision of public transport services.

Road infrastructure development

While there are unquestionably many benefits to the development of road infrastructure for poor rural communities, governments have limited budgets and can thus only provide limited new development. The decision-making process for when and where to build new road infrastructure and/or upgrade existing roads needs to incorporate two components: maximising the number of people benefiting from the new roads, and minimising the costs to users (Shrestha et al., 2014). Projects must also incorporate sustainability objectives throughout the development process including during decision-making and project design. Faiz et al. (2012) state that:

> to be sustainable road infrastructure projects must firstly contribute to and enhance rural livelihoods and liveability, and secondly their planning and design (as well as construction and maintenance) must be context sensitive to ensure a balance among economic, social and environmental objectives that are reflective of community values, aspirations and needs.

While most governments acknowledge this in theory, their true adherence to sustainable decision-making has been questioned. Malvestio et al. (2018) for example, investigated transport decision-making in Brazil. They found that while the government paid lip service to the three pillars of sustainability, in reality their decision-making and road design had been based purely on economics with environmental and social considerations tacked on as an afterthought. Their environmental considerations only accounted for compliance with greenhouse gas emission minimisation, which is a statutory obligation, and failed to consider other important issues such as biodiversity. Social impacts were incorporated during late stages of the project due to the strong influence of non-profit organisations and other activist groups. If these groups had been consulted at the design stage the overall process could have been more economical than spending time and money coming up with a design only to have to go back and make adjustments for social and environmental considerations at a later stage.

There are various examples in the literature of road infrastructure development failing to achieve these goals. For example, in many African nations road infrastructure development prioritises connecting rural communities with urban centres, rather than connecting adjacent rural communities. From the perspective of farming communities, looking for a market for their produce, there is some logic to this, in that a densely populated urban area would theoretically provide a bigger market. On the other hand, if multiple rural communities are all competing to sell their produce in the same market place, the competition may keep prices low reducing profits. Calmette and Kilkenny (2012) modelled various economic scenarios and showed that sharing produce between rural communities can in some cases have more significant economic benefits. One area they focused on was the situation where one of the rural communities experiences a poor harvest, and thus is forced to rely on produce from external sources to make up the shortfall. In a developed country this situation presents an opportunity for the non-drought affected area to increase sales by supplying the region normally covered by the drought affected area. In many developing countries particularly sub-Saharan Africa, this is not the case because the cost of transport is an estimated 2.5 times higher than anywhere else (Mwase, 2003). In the present example, produce must be transported to an urban centre, stored briefly at cost, and then transported back out to another rural area. The storage plus transport costs were shown to outweigh the value of the produce, making it economically unviable for adjacent rural communities to supply each other. This problem is somewhat exacerbated by food aid, which provides a cheap alternative to locally grown produce. The authors suggest that money invested in food aid by developed nations would be better spent improving transport infrastructure, allowing trade between adjacent rural communities. In the long run this would make the region more sustainable as it would immediately increase market opportunities, as well as allowing adjacent communities to pick up the shortfall when one or the other experiences a poor harvest. This is especially important now as occurrences of extreme weather events have been increasing in frequency and length and are projected to continue to do so. It would also benefit the environment as it would reduce greenhouse gas emissions associated with transporting goods, especially compared to food aid, which is supplied by plane from foreign countries.

One of the main reasons for this lack of foresight into the most appropriate road infrastructure development is the failure to consult with the rural community about their transport requirements. Many other studies found a similar failure to achieve development goals through road infrastructure due to lack of public consultation. Asomani-Boateng et al. (2015) investigated the long-term results of a road infrastructure upgrade programme in Ghana. While there were some successes, for example average travel times to schools decreased improving educational attainment opportunities for many children, the authors noted that the scheme could have been implemented more efficiently. The campaign focused on building highly engineered asphalt roads suitable for high flows of motorised vehicles, when a significant number of the roads had low vehicle movements and were mainly used by pedestrians, cyclists or animal-based transport. This was a result of the lack of public consultation at the design stage. The local residents are the ones set to benefit from the implementation of the roads. They know where they need to go and what means of transport they will use to get there. Therefore their input into the design is invaluable.

Road maintenance

Worldwide there is an estimated 33.8 million km of classified public roads, around 0.23 km of road per km^2 of land area (Faiz et al., 2012). If all of these roads were left to fall into disrepair Faiz et al. (2012) estimate that the cost of replacement would amount to 7.6 trillion USD. Thus, roads in themselves are a huge asset worth preserving. Poor road maintenance will also negatively impact the economy in other ways. Where roads become impassable economic activity will be halted. For example, farmers will not be able to transport their produce to markets to sell, hindering their ability to generate an income. Where roads are still in use, but in poor condition vehicle fuel use and tyre wear will increase, leading to increased cost of vehicle repair and maintenance. Produce and goods being transported to markets will also be at higher risk of being damaged. Steyn et al. (2015) measured the impact of vibrations in a typical heavy goods vehicle on tomatoes. They tested the tomatoes for vehicles travelling on various road types at varying speeds. They found that at least some of the tomatoes were damaged by the vibrations in all scenarios, which in some instances, would have resulted in tomatoes being sold at reduced price, while others were unsaleable and had to be discarded. They estimated the loss of income to be 8 per cent compared to selling all tomatoes at full price. This small laboratory experiment was based on typical Californian roads. The experimental procedure could be replicated in other regions; however it seems likely that impacts due to a poorly maintained road in a developing country are likely to be more severe than one in a developed region like California.

Despite the economic benefits of routine road maintenance, it is still severely lacking in many developing nations. Pinard et al. (2016) notes that of the 2 million km of roads constructed in Africa between the 1950s and 1980, around 33 per cent had fallen into disrepair by 2000. Mwase (1996) reports that in Tanzania, where 100 per cent of roads were in good condition in the 1970s, only 30 per cent remained in good condition by 1990. Many had become impassable, particularly during the rainy season regularly cutting off rural areas from obtaining supplies or taking their goods to markets. Asian Development Bank (2003) reports on a similar lack of road maintenance in six Asian countries, Kyrgyzstan, Laos, Pakistan, Philippines, Uzbekistan and Vietnam. Thus, poor road maintenance is a widespread problem spanning multiple regions on multiple continents.

The Asian Development Bank (2003) investigated some of the reasons why road maintenance is neglected in developing nations. One of the main problems is that the government has a single budget stream for the transportation department, and therefore is often deciding whether to spend money on building a new road or maintaining an existing one. From the government's perspective:

- Building a new road attracts a lot of positive publicity.
- The cost of replacing a road after 20 years is similar to the sum total of annual maintenance and repair costs over the same time period (if you ignore the negative economic impacts experienced by the road users).
- Spending money to maintain a road in one village, while other villages have no roads could be perceived as unfair. The government wants to appear fair, by spending money on more villages.
- Identifying minor damage that requires maintenance is not always easy. Many governments take the attitude "if it ain't broke don't fix it", and thus will not spend money on repairs unless they can easily see the damage. They fail to comprehend that small routine maintenance could be performed cheaply and prevent the need for more complex and expensive repairs later on.
- Inspecting roads to look for minor damage requires someone with formal training. In many regions only engineers have this training, however, it is tedious work and engineers would rather be involved in the construction of a new road which would bring them greater prestige.

So how can these obstacles be overcome?

Road maintenance should be budgeted for at the road design and construction phase, so that there is no competition for the money later on. An important way to get governments on board with this approach is to highlight the broader economic impacts of poor roads. In many regions, road construction is funded by an external party such as an NGO. These organisations should include money for road maintenance, with the condition that it cannot be spent for another purpose. Operating a "user pays" system has also been found to be effective in many regions. This can take the form of taxes on vehicle ownership and fuels, or toll roads where appropriate. Where taxes on fuels are applied lengths should be taken to ensure only road users are being taxed. Many small rural communities rely on diesel powered generators to provide electricity, and should be exempt from fuel tax. User pays strategies must be stringently regulated to ensure that the funds are only available for use for road maintenance and the government cannot divert them to other purposes.

Another way to make road maintenance more economically efficient is ensuring roads are fit-for-purpose. As we mentioned in the road infrastructure development section, governments prefer to build high tech asphalt roads, to boost their profile. However, these roads are not only more expensive to construct, but also are more difficult and expensive to maintain. In areas where roads have low motorised vehicle movements, and are more frequented by pedestrians, cyclists and so on, a gravel or dirt road may provide the more sustainable solution once the budget is extended to include maintenance (Asomani-Boateng et al., 2015, Asian Development Bank, 2003).

Localising road maintenance has also been shown to be an effective strategy. Hiring an engineer to inspect the roads periodically will be expensive and will mean that minor damage will not be picked straight away. Training members of the local community, who are using the road every day, to identify damage that requires maintenance would be cheaper and would ensure damage was picked up early on, allowing simple maintenance to be performed and preventing the need for more expensive and complex repairs later on. The maintenance would ideally also be able to be performed by someone local using locally available materials. An example of a successful scheme is the use of "do-nou", basically a bag filled with dirt. The bags are made from a material with high tensile strength that can withstand frequent vehicle movements. Testing by Fukubayashi and Kimura (2014) found that bags used for produce storage that are commonly found in rural areas were suitable for this purpose. Training locals in how to use the bags for road maintenance is simple. The first step is sizing the bags so that once filled with dirt they can be easily lifted by a person. Once the bags have been appropriately sized and filled with soil, they can be placed at points in the road that need reinforcing. The method was originally trialled in Kenya and the authors investigated the success of the programme up to two years after initial implementation. They found that not only had the repairs been successful, but also that the programme had expanded. At first many locals were cynical of the method, but after people noticed that the road repairs were holding up under traffic, they became interested in learning the technique as well and more and more residents began to use the bags to repair roads in their local areas. The technique is now used in 13 countries throughout Asia and Africa and is continuing to expand.

In research conducted by one of the authors on maintenance of rural road in India, it was found that heavy top-down directives in public governance is a problem in timely and effective maintenance of rural roads. A suggestion was made for promoting a community centric approach for maintaining rural roads as opposed to the three tier governance structures involving state, district and panchayat level bureaucracy as prevails in most Indian states. In the alternative approach, a hybrid funding model is proposed for raising funds from both public agencies (such as district level funding allocation) and private sources (such as philanthropic donations or NGOs) that is controlled by a local committee for necessary scope of work within their jurisdiction. The local committee would preferably be comprised of technical people from the community and would be responsible for overseeing the expenditure and ensuring quality of work done by the subcontractors or any other qualified agencies. The model is expected to be piloted in the state of Assam as part of the current Smart Villages programme.

Beyond roads: completing the picture of transport infrastructure

While the provision and maintenance of road infrastructure is an important component of connecting rural communities with the outside world, it is not the complete picture. In many developing economies, only a minority of people own a motorised vehicle. For example, the University of Melbourne surveyed residents of Majuli Island, Assam, India on the modes of transport they used to get to work as shown in Figure 9.2. It can be seen that more than three quarters of people commute to work by walking, while around one quarter use either bicycle or motorcycle/scooter. Less than 1 per cent of people commute by car. Alveano-Aguerrebere et al. (2017) surveyed transport modes of residents of Morelia, Mexico and the distribution is in shown in Figure 9.3. Walking is again the most common mode, followed by public transport. Morelia is a more urban area thus the high use of public transport is most likely a reflection of its greater availability. Similarly, Johnston (2007) surveyed communities in West Java, Indonesia and found that less than 10 per cent of households owned a motorised vehicle. Walking was the most common mode of transport accounting for around 90 per cent. Henseler and Maisonnave (2018) report that in South Africa less than one quarter of the population owns a car, and those that do generally have higher socio-economic status. These facts are often overlooked during the design and construction of new road infrastructure, which is focused solely on providing infrastructure for motorised vehicles, resulting in them being unsuitable for the majority of users.

One of the consequences of this, as seen in Chapter 10, is that road traffic accidents are still one of the leading causes of death in low and middle-income countries. Data from WHO show the number of deaths from road collision injury in developing countries is not just high but is increasing. For example, Jagnoor et al. (2018) report that road collision death rates in India increased

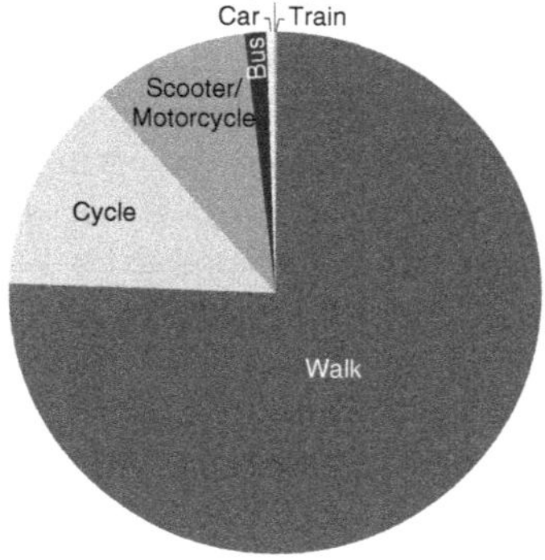

Figure 9.2 Distribution of modes of transport used to get to work by Majuli Island residents

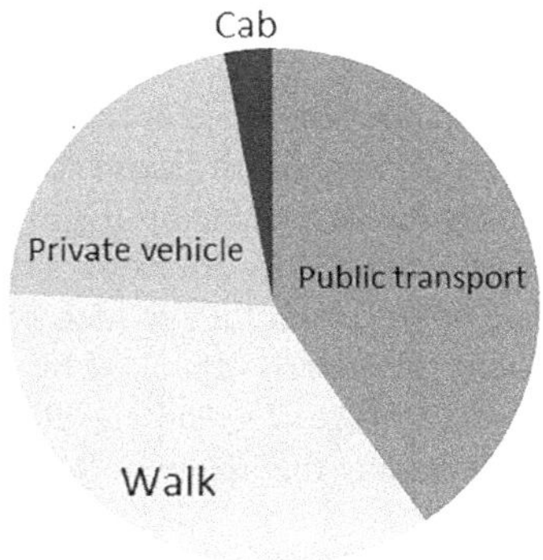

Figure 9.3 Frequency of use of mode of transport in Morelia, Mexico (adapted from Alveano-Aguerrebere et al., 2017)

by 8 per cent per year between 2000 and 2006. Various studies into road traffic victims show that non-motorists are disproportionately affected. In Kenya, Wells (2007) found that only 10 per cent of road collision fatalities were car drivers, most were pedestrians or cyclists. Alveano-Aguerrebere et al. (2017) report that pedestrians and cyclists account for 40 per cent of road deaths in Mexico. McAndrews et al. (2017) similarly found pedestrians and cyclists were at higher risk of injury in the US state of Wisconsin. WHO projects that without significant intervention this death toll will continue to increase as rapid development in low to lower middle income countries results in rapid increases in car ownership, without any associated increase in road safety initiatives. Therefore, it is clear that transport infrastructure must cater to multiple forms of transport, especially non-motorised.

Surveying the local communities about their transport requirements during the design stage of a road infrastructure project is an important step in determining what modes of transport the road needs to cater for. The most common road feature that can improve safety for non-motorised forms of transport is a separate path. In some areas this is simply marked with a line with sign posts indicating that it is not to be accessed by cars. However, Alveano-Aguerrebere et al. (2017) surveyed residents of Morelia, Mexico on whether they thought that this would be sufficient in their region. While most respondents agreed a separate lane was necessary, they did not believe that vehicles would stay out of the lane unless there was a physical barrier to prevent them from entering it. In many developing regions road rules are lacking and where they do exist they are poorly enforced, meaning drivers are not used to having to obey signs (Wells, 2007). Yannis et al. (2011) explored patterns of road collision incidents over time in various European countries. Their main finding was that road tolls increased in line with vehicle ownership to a point but then eventually plateaued. The reasons for this were estimated to be improved safety features, such as separate cycle lanes, and expanded provision of public transport. Therefore, incorporating separate paths for non-motorised transport modes with a physical barrier should be a part of any new road infrastructure design.

A final consideration is the relationship between transport mode and trip length. Table 9.1 shows mean distance by mode of transport for Majuli Island residents. We can see that while walking and cycling are used for frequent short trips, longer journeys tend to use buses or trains. This finding is echoed by Johnston's (2007) survey in West Java, Indonesia which found mean distance of a bus journey was around 20 km, while shorter distances were more frequently travelled by walking or motorcycle taxi. The author also noted that villages located closer to bus stops had a lot fewer motorcycle taxi journeys, indicating that many of these were simply to access bus services.

Thus, buses and other public transport services are relied upon for longer journeys. As younger people continue to migrate to urban areas en masse, leaving behind an aged rural population, the ability to access services via public transport is crucial to survival (Ryser & Halseth, 2012). Yet, public transport services are often woefully inadequate. Lucas (2011) investigated public transport services in South Africa, and found that they were expensive, unreliable and often unsafe, exposing users to accidents, crime, noise and pollution. Children travelling to school were noted to be especially at risk. Due to a lack of options one student had to decide whether to walk several miles to school or take a motorcycle taxi and forgo having lunch that day. Other children who were walking up to seven miles to attend school were found to be vulnerable to violent crimes or theft or were recruited to commit crimes by gangs on their way to school. The article cited similar difficulties in accessing health services, or even getting to work. The opportunity for young people in urban areas to visit their rural homes was also noted as a priority as this was their main way of staying in contact with family and friends.

Ascertaining funding for public transport services remains the greatest obstacle to their provision. In the case of South Africa, fuels are currently subsidised by the government. Given the fact that only a quarter of the population own a car, and that the majority of these are in the higher socio-economic classes,

Table 9.1 Mean distance travelled for each mode of transport by residents of Majuli Island

Mode of transport	*Percentage of respondents*	*Number of respondents*	*Mean distance travelled (km)*
Walk	75.5	1825	8
Cycle	13.1	317	6
Scooter/Motorcycle	9.4	228	15
Bus	1.3	32	29
Car	0.5	11	5
Rickshaw	0.0	1	3
Train	0.1	3	250

diverting this subsidy to public transport has been suggested as a logical way to provide transport infrastructure that caters to the majority of the population (Henseler & Maisonnave, 2018). This suggestion is being made in the wake of a recent drop in the price of fuels, meaning removing the subsidy now would not result in a significant increased cost to fuel users. The fuel subsidy also goes against South Africa's commitment to greenhouse gas emissions reductions, whereas diverting the funds to improve public transport services would not only serve this commitment, but also aid attainment of Sustainable Development Goals. Economic modelling by Henseler and Maisonnave (2018) of various scenarios of diverting fuel subsidies to public transport all resulted in poorer households being able to increase their income and spend less on travel. This led to an overall improvement of the economy as small businesses were able to increase number of employees etc. Where there is no obvious financing available for improving public transport infrastructure, it should instead form a part of the design stage of transport infrastructure development. The transport infrastructure development should be budgeting not just for the building of roads, but for their maintenance and provision of public transport services. This is the only way that they can effect real change in the lives of poor rural communities.

Recommendations for best practice

The decision-making process for transport infrastructure development should aim to impact as many people as possible, while adhering to sustainability objectives. The process should incorporate consultation with potential future road users as these are the only people who know where they need to go and what means of transport they will use to get there. These are the key considerations which should inform the design and development of new road infrastructure.

Regular road maintenance is essential to provide long-term transport connections to villages and to ensure money spent implementing road infrastructure is not wasted. Thus, budgeting for maintenance should form part of the road design phase. This could include setting funds aside from the outset, or implementing a tax on road users. Localising road maintenance has also been shown to be an effective way to prevent roads falling into disrepair. Training people who use the road every day to identify the need for maintenance and perform basic repairs ensures roads do not fall into disrepair and minimises the costs of road maintenance. If locals can perform the maintenance with locally available materials this will be even better as it will ensure maintenance is performed as early as possible.

Motorised vehicles are often a small proportion of rural road users Therefore, road infrastructure should include features that cater for non-motorised forms of transport such as pedestrians and cyclists. This will be essential for reducing the high number of road collision related deaths and injuries that are currently occurring in many developing regions.

Many rural residents rely on public transport services such as buses or motorcycle taxis to access everything from places of employment to schools and healthcare facilities. Improving the reliability and safety of these services needs to be prioritised to effect real change in rural communities. Public transport provision must form a part of the design phase of transport infrastructure to ensure it is adequately budgeted for.

References

Alveano-Aguerrebere, I., Ayvar-Campos, F. J., Farvid, M. & Lusk, A. 2017. Bicycle facilities that address safety, crime, and economic development: Perceptions from Morelia, Mexico. *International Journal of Environmental Research and Public Health*, 15, 22.

Asian Development Bank 2003. *Road funds and road maintenance: An Asian perspective.* Manila, Philippines: Asian Development Bank.

Asomani-Boateng, R., Fricano, R. J. & Adarkwa, F. 2015. Assessing the socio-economic impacts of rural road improvements in Ghana: A case study of Transport Sector Program Support (II). *Case Studies on Transport Policy*, 3, 355–366.

Calmette, M.-F. & Kilkenny, M. 2012. Rural roads versus African famines. *The Annals of Regional Science*, 49, 373–396.

Faiz, A., Faiz, A., Wang, W. & Bennett, C. 2012. Sustainable rural roads for livelihoods and livability. *Procedia – Social and Behavioral Sciences*, 53, 1–8.

Fukubayashi, Y. & Kimura, M. 2014. Improvement of rural access roads in developing countries with initiative for self-reliance of communities. *Soils and Foundations*, 54, 23–35.

Henseler, M. & Maisonnave, H. 2018. Low world oil prices: A chance to reform fuel subsidies and promote public transport? A case study for South Africa. *Transportation Research Part A*, 108, 45–62.

Jagnoor, J., Prinja, S., Lakshmi, P. V. M., Aggarwal, S., Gabbe, B. & Ivers, R. Q. 2018. The impact of road traffic injury in North India: A mixed-methods study protocol. *BMJ Open*, 5, 5.

Johnston, D. C. 2007. These roads were made for walking? The nature and use of rural public transport services in Garut Regency, West Java, Indonesia. *Singapore Journal of Tropical Geography*, 28, 171–187.

Kilkenny, M. 2010. Urban/regional economics and rural development. *Journal of Regional Science*, 50, 449–470.

Lucas, K. 2011. Making the connections between transport disadvantage and the social exclusion of low income populations in the Tshwane Region of South Africa. *Journal of Transport Geography*, 19, 1320–1334.

Malvestio, A. C., Fischer, T. B. & Montaño, M. 2018. The consideration of environmental and social issues in transport policy, plan and programme making in Brazil: A systems analysis. *Journal of Cleaner Production*, 179, 674–689.

McAndrews, C., Beyer, K., Guse, C. & Layde, P. 2017. Linking transportation and population health to reduce racial and ethnic disparities in transportation injury: Implications for practice and policy. *International Journal of Sustainable Transportation*, 11, 197–205.

Mwase, N. 1996. Developing an environment-friendly transport system in Tanzania: Some policy considerations. *Transport Reviews*, 16, 145–156.

Mwase, N. 2003. The liberalisation, de-regulation and privatisation of the transport sector in sub-Saharan Africa: Experiences, challenges and opportunities. *Journal of African Economies*, 12, ii153–ii192.

Pinard, M. I., Newport, S. J. & Rijn, J. V. 2016. Addressing the road maintenance challenge in Africa: What can we do to solve this continuing problem? *International Conference on Transport and Road Research*. Whitesands Hotel, Mombasa, Kenya.

Rönkkö, E., Luusa, A., Aarrevaara, E., Herneoja, A. & Muilu, T. 2017. New resource-wise planning strategies for smart urban-rural development in Finland. *Systems*, 5, 12.

Ryser, L. & Halseth, G. 2012. Resolving mobility constraints impeding rural seniors' access to regionalized services. *Journal of Aging and Social Policy*, 24, 328–344.

Shrestha, J. K., Benta, A., Lopes, R. B. & Lopes, N. 2014. A multi-objective analysis of a rural road network problem in the hilly regions of Nepal. *Transportation Research Part A*, 64, 43–53.

Steyn, W. J. V., Nokes, B., Plessis, L. D., Agacer, R., Burmas, N. & Popescu, L. 2015. Evaluation of the effect of rural road condition on agricultural produce transportation. *Transportation Research Record*, 2473, 33–41.

Wells, P. 2007. Deaths and injuries from car accidents: An intractable problem? *Journal of Cleaner Production*, 15, 1116–1121.

Yannis, G., Antoniou, C., Papadimitriou, E. & Katsochis, D. 2011. When may road fatalities start to decrease? *Journal of Safety Research*, 42, 17–25.

10 Health and well-being

Introduction

Three of the Millennium Development Goals are directly related to health: Goal 4: Reduce Child Mortality, Goal 5: Improve Maternal Health and Goal 6: Combat HIV/AIDS, Malaria and Other Diseases. However, achieving any of the goals would lead to improved health outcomes.

- *Goal 1*: Eradicate Extreme Poverty and Hunger – malnutrition hinders both physical and cognitive development of children. Malnourished people's immune systems are weaker making them more vulnerable to infectious diseases. Inability to afford basic provisions leads desperate people to ingest food and water that are contaminated.
- *Goal 2*: Achieve Universal Primary Education – education programmes include promotion of basic hygiene practices such as washing hands, preparing food safely, understanding of what causes diseases and how to prevent them spreading.
- *Goal 3*: Promote Gender Equality and Empower Women – Women often take on the primary care role for children. Children in rural areas are 1.7 times more likely to die before reaching the age of five than their urban counterparts. This is mainly a product of the low educational attainment of women in these areas. Children of mothers with secondary or higher education are almost three times more likely to survive beyond the age of five than those of mothers with no education (United Nations, 2015).
- *Goal 7*: Ensure environmental sustainability will also impact health. In fact, much of the environmental legislation in force throughout the world today was motivated by human health related incidents. The UK Clean Air Act was developed in response to the severe "pea soup" smog incident in 1952. During the incident 4000 deaths and 100,000 instances of illness were attributed to the smog. After the incident elevated mortality levels continued to be observed. At the time these deaths were attributed to an outbreak of influenza. However, a retrospective analysis (Bell & Davis, 2004) showed that this would have been an influenza epidemic three times larger than the most severe epidemic on record, which is extremely unlikely. The elevated level of pollution during the smog event is much more likely to

> have been the cause. Thus, in 1956 the Clean Air Act was initiated. In the US controls on the use of pesticides was prompted by a spate of paralysis symptoms caused by "Ginger Jake", an alcoholic beverage made from fermented Jamaican ginger. This product was sold as a medical tonic during the prohibition era, to bypass the ban on alcohol. The use of pesticides on the product was heavy, and manufacturers exacerbated its use, as it helped to confound alcohol testing (Davis, 2014).

A similar argument could also be made for many of the Sustainable Development Goals. Maintaining good health, particularly during childhood is the key to allowing people to achieve their full potential, which in turn is the key to maximising their contribution to economic development.

A good way to understand how healthcare problems vary between regions is to look at the variation in leading causes of deaths. Figure 10.1 shows the top ten causes of death globally for 2015, which includes more than 50 per cent of the total number of recorded deaths. Ischaemic heart disease and stroke are the most common causes, followed by respiratory and lung related illness.

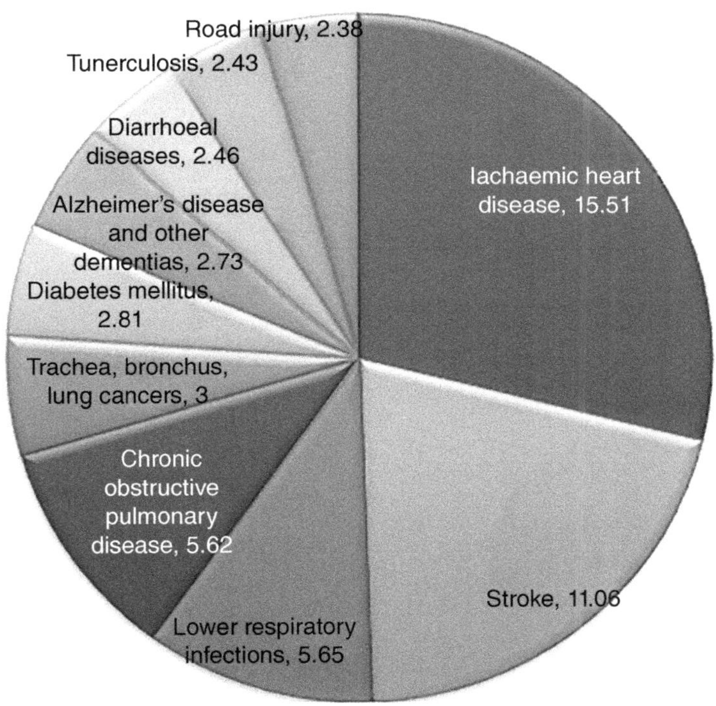

Figure 10.1 Top ten causes of death globally for 2015, showing percentage values (adapted from World Health Organization, 2018a)

Figure 10.2 shows the top ten causes of death separated into World Bank income ranges. The first notable difference is the prevalence of death due to diarrhoeal diseases in low and lower-middle income countries accounting for 6.7 per cent and 4.1 per cent of deaths respectively, while not featuring in the top ten of upper-middle and high-income countries at all. As discussed in Chapter 8, one leading cause of these is coming into contact with faecal matter, which can be prevented by the use of improved sanitation facilities. The other leading cause is foodborne illness, which will be discussed in more detail in the first section of this chapter.

Two of the leading causes of death in low income countries do not appear in the top ten of any other income group. These are malaria and HIV/AIDS, two communicable diseases that are preventable and treatable. Tuberculosis also features prominently in both low and lower-middle income countries despite being preventable and curable. The second section of this chapter will discuss prevention and treatments of these three communicable diseases.

Deaths due to pregnancy or complications during birth are another cause of death dominant in the low and lower-middle income countries that falls outside of the top ten causes of death in higher income regions. The presence of a qualified

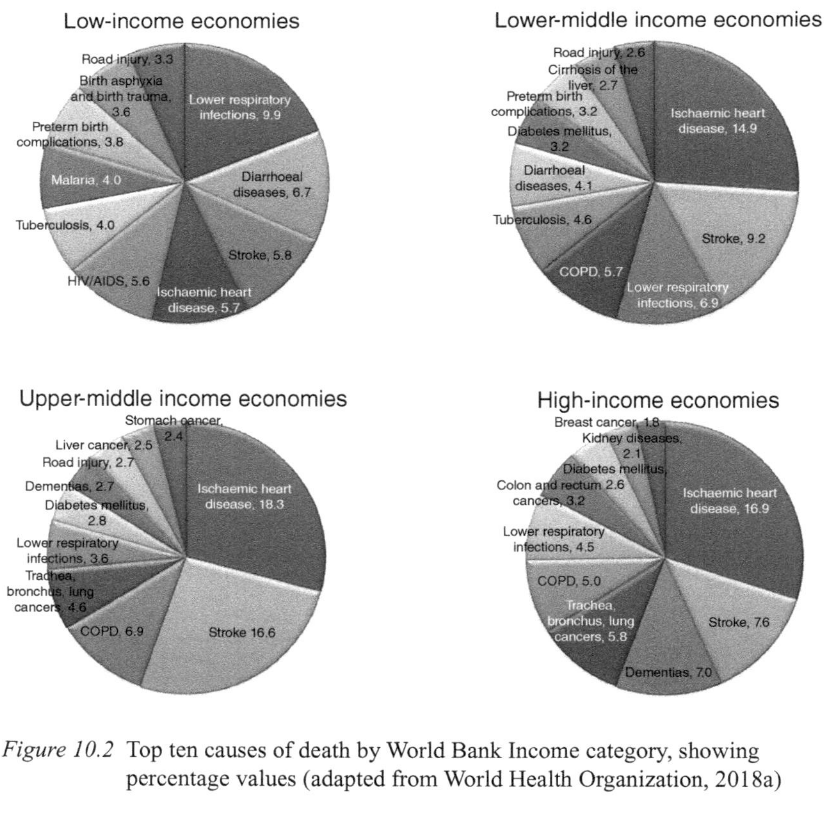

Figure 10.2 Top ten causes of death by World Bank Income category, showing percentage values (adapted from World Health Organization, 2018a)

healthcare professional during birth could significantly reduce deaths in this area. The lack of medical assistance during a birth has also been noted as a mark of the urban-rural divide, where only 56 per cent of births in rural areas are attended by skilled health professionals, compared to 87 per cent in urban areas (United Nations, 2015). In general, better access to healthcare could reduce incidence of diseases, while others could be treated more effectively preventing them leading to mortality or lifelong disability. Therefore, the second section of this chapter will discuss affordable ways to improve access to healthcare professionals.

Death due to road traffic collisions appears in the top ten of all but the high-income regions. In Chapter 9 we already discussed the need for improved road maintenance to enhance safety not just of drivers, but other road users, especially pedestrians and cyclists. Another reason for the high incidence of death in lower income regions is lack of or inadequate emergency response. This holds true for not just road collisions, but many other forms of trauma, which are more common in some regions. For example in South America, the leading cause of trauma is violence (Poggetti, 2008), and this is likely to be similar in other locations with social inequality and political instability. Thus, the last section of this chapter will look at improving provision of emergency response.

Foodborne illness

Foodborne illness is a worldwide phenomenon but is particularly prevalent in developing economies and rural areas. The WHO assessed the burden of global foodborne illness for 2010 and estimated that 600 million cases of illness and 420,000 deaths occurred (World Health Organization, 2015). Of these 550 million instances of illness and 230,000 deaths were due to diarrhoeal disease. Children under the age of five were disproportionately affected, accounting for 40 per cent of foodborne illnesses, even though they only account for 9 per cent of the population. Many of the affected children die and of those that survive many suffer lifelong disabilities hindering their ability to reach their full developmental potential, and contribute meaningfully to economic development.

Foodborne illness can have multiple negative impacts on the ability of people to contribute to economic development:

- It can lead to death or disability reducing the labour force available.
- Knowledge of an outbreak of foodborne illness will lead to restrictions on the saleability of agricultural produce, particularly where this is exported. Consumers often react to knowledge of a foodborne illness outbreak by refusing to buy produce from a whole country, even if the outbreak only occurred in a single region.
- Disease outbreaks will also deter tourists from visiting a region, leading to a substantial loss of potential income.

The disparity in the prevalence of foodborne disease between different regions of the world indicates that these diseases are preventable in many instances and interventions to reduce foodborne illness should be a priority.

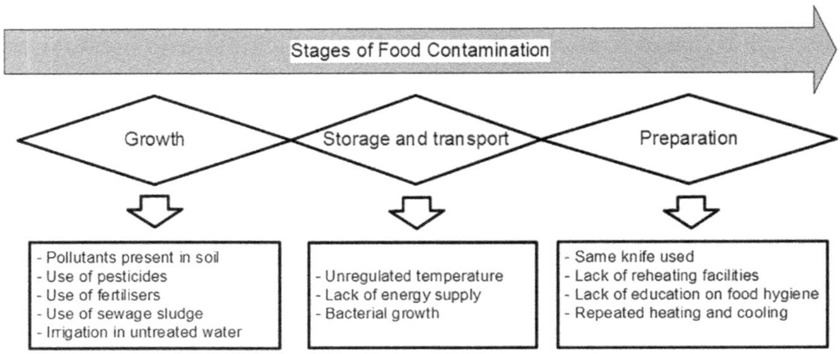

Figure 10.3 Stages of food contamination

Food can become contaminated at many different points during its life cycle as illustrated in Figure 10.3.

Thus, in the rural community there are many factors contributing to the potential for foodborne illness outbreaks. Many of the topics covered previously in this book elaborate on these issues. For example, lack of access to energy means no refrigeration for safe food storage, and inadequate cooking facilities for safe food preparation. Lack of fresh water makes it difficult to wash fresh produce to remove pesticides and fertilisers. The inability to test soil quality and poor use of pesticides and fertilisers can lead to accumulation of not only pathogens but also other potentially toxic elements such as heavy metals in food. This highlights the fact that improved economic development will lead to improved health outcomes.

Prevention and treatment of communicable diseases

Malaria

Malaria is transmitted to humans through the bites of the female *Anopheles* mosquito. It is mainly found in sub-Saharan Africa, where 90 per cent of cases, and 91 per cent of malaria related deaths occurred in 2016, however, cases are also noted in South East Asia, Eastern Mediterranean, Western Pacific and the Americas regions (World Health Organization, 2018c). Contraction of malaria does not necessarily result in the development of symptoms and partial immunity can be built up over time. Some members of the community are more at risk of developing symptoms than others including young children, pregnant women, people with HIV/AIDS, international visitors, travellers and mobile populations who would not have built up any immunity.

The most effective way to prevent the spread of malaria is the use of insecticides to control mosquitoes. The most commonly employed methods are indoor spraying, and the use of insecticide treated nets during sleep. Since 2000 various campaigns such as the US President's Malaria Initiative and Roll Back Malaria have seen the use of these two methods, mostly by way of pyrethroid insecticides, ramped up. As a result, malaria related deaths have halved, and incidence of disease have been reduced by over one third (Ranson & Lissenden, 2016). Unfortunately, in more recent years, incidence of malaria appears to be taking an upturn and evidence is starting to accumulate that mosquitoes in many of these regions have begun developing resistance to pyrethroid insecticides. While the use of nets still has some benefit, by providing a physical barrier, the continued use of this chemical in indoor spraying is now becoming redundant. It is clear that new insecticides that target mosquitoes must be developed; this development should take heed of the fact that mosquitoes will likely develop resistance to them once their use becomes frequent and widespread. Ranson and Lissenden (2016) state that researchers should aim to develop at least three different new chemical agents, rather than aiming for a single insecticide. Using these multiple insecticides in rotation will prevent or at least significantly delay mosquitoes developing resistance, providing a longer-term solution.

Benelli and Beier (2017) suggest an integrated vector management (IVM) approach should be employed, similar to the integrated pest management approach described in Chapter 4. An IVM system would continue to employ indoor spraying and nets, while also incorporating other novel approaches to mosquito control. For example, the insertion of plastic tubes between the roof and walls of houses (where there is commonly a gap in typical African house design) that have been treated with an insecticide. It has been noted that mosquitoes often enter a house through this gap so would be an effective component of an overall IVM strategy. They note however, that before any new methods are used, their effectiveness, and impact on other components of the ecosystem must be assessed to prevent disaster. They also suggest adding species of fish that consume mosquito larvae to nearby waterways could be effective, although there could be unintended consequences to the overall ecosystem which should be considered. Two of the key components of an IVM are monitoring programmes to look for early indications of mosquito resistance to pesticides, so that alternative vector management techniques can be employed. The monitoring should also note how malaria incidences are affected by changes in climate to help us understand how mosquitoes are responding to these. Finally, developing a deeper understanding of mosquito behaviours will be important for targeting prevention strategies more effectively.

Malaria is treatable, and the formation of symptoms can be prevented through the use of medicine. The discovery of Artemisinin, the active ingredient in many malarial drugs, was a huge breakthrough in malaria related mortality reduction (Wells et al., 2015). It has been used to treat hundreds of millions of malaria-infected patients. Three obstacles to the effectiveness of treatment exist. Firstly, the market is plagued with counterfeit medicines. Secondly, most treatments involve multiple doses to be taken over a period of time (usually three days). However, once symptoms subside many people decide to save the remaining medication in case another family member is infected later. This partial treatment means not only that the person has not effectively removed the malaria parasite from their system, but it also encourages the emergence of resistant strains of malaria. Thirdly, the parasite is developing resistance to artemisinin. Similar to the issues surrounding pyrethroids, the use of a singular compound has promoted the development of resistance. Therefore, it is essential that continual research into new compounds that can treat malarial infections continue and that multiple compounds be used in rotation. Another problem with artemisinin is that while it is fast acting it does not prevent reinfection. Thus, it is not the most appropriate treatment for people experiencing frequent exposure, particularly, children in Africa who are bitten by mosquitoes nearly every night.

The development of a vaccine could potentially be a much more effective long-term solution to the malaria problem (Hoffman et al., 2015). The complication is that malaria is caused by a parasite, rather than a virus or bacteria. Viral and bacterial infections have a single stage in their life cycle, therefore, the immune system only needs to develop a single response mechanism to fend off these infections. Parasites on the other hand have multiple stages in their life cycle; therefore the immune system would need to develop multiple response mechanisms to fend off the parasites at these different stages (Hoffman et al., 2015). Having said that, there are many malarial vaccines currently under development, some at quite advanced stages, therefore there may be some hope for this in the near future.

HIV/AIDS

The human immunodeficiency virus (HIV) is an infection that weakens the immune system. At its most advanced stage it results in autoimmune deficiency syndrome (AIDS) anywhere between two and 15 years after initial infection. The onset of AIDS for HIV infected people can be significantly delayed by antiretroviral (ARV) medication. AIDS in itself is not the cause of death, but due to the weakened state of the immune system other diseases are more easily contracted, and symptoms are experienced at an advanced rate. Africa is the most affected region, accounting for nearly 70 per cent of people living with HIV/AIDS in 2016 (World Health Organization, 2018b).

HIV/AIDS is most commonly spread through sexual intercourse or intravenous drug use. The most well-known way to prevent infection is the use of condoms, which have been shown to be 85 per cent effective at preventing transmission from sexual intercourse, and sterilising and/or not sharing needles/syringes. In 2011, trials showed that the use of ARV medication also reduced the risk of an infected person transmitting the disease by 96 per cent. The use of ARV medication by HIV-negative people in a high exposure risk situation has also been shown to prevent contraction of HIV. These include partners of infected persons, or children of infected mothers, particularly during breastfeeding. This is known as pre-exposure prophylaxis treatment (PrEP). The use of ARV medication by pregnant and lactating women has been shown to prevent spread of infection to babies.

For people who have been exposed to HIV a 28 day course of ARV medication has been shown to significantly prevent development of the disease. This is known as post-exposure prophylaxis (PEP). It usually also involves counselling, testing and follow-up care.

There are three main hindrances to the eradication of HIV: First is the availability of ARV medication. In 2016 only 53 per cent of HIV infected people were receiving ARV treatment. Children especially lacked access to medication, at only 43 per cent, while pregnant women received better treatment at 75 per cent.

Second is ensuring patients adhere to the medication regime. Doses need to be taken multiple times per day, and prescriptions are often only filled for a month's worth at a time. Attending a monthly clinical appointment can be burdensome for many people, especially those in rural areas who may be required to travel long distances. Clinics are often overcrowded, and patients usually have to wait a long time to see the doctor. Ensuring patients understand the mechanism of their medication is also important. In South Sudan for example, the Star Support Group who work with HIV infected people found many believed the medication should only be taken after AIDS symptoms were developed and did not realise they should be used to prevent spread of the disease as much as they are to suppress the symptoms (Personal Communication, G. Ellul).

The final issue is that because the early stages of HIV are symptomless, many people are likely to be unaware that they have the virus. Encouraging people to get tested, especially if they are in a high-risk situation is crucial. The use of self-administered testing kits has been trialled and could be a significant boost to encouraging testing in those who are embarrassed or scared to seek testing.

Early global commitment to tackling the HIV epidemic saw huge amounts of funding to low and lower-middle income countries, allowing the expansion of ARV treatment to an unprecedented number of patients (Duncombe et al., 2015). However, this funding is now beginning to taper off, and is predicted to begin a downturn in the near future as other global commitments come to the fore. Most of the funding in low and middle-income countries went to establishing a clinic based system for HIV treatment, regardless of the health status of patients. As a consequence, medical resources are becoming stretched by the volume of HIV patients and frequency of appointments. Patients are also overly burdened by the need to attend a clinic and wait hours for an appointment simply to refill a prescription. As HIV is not curable, patients will be required to continue taking the medication for the rest of their lives, and this situation is not going to be feasible for many, especially those in rural areas, who may need to travel long distances to attend clinics. Bemelmans et al. (2014) note that in the Democratic Republic of the Congo, the combined cost of transport to a clinic, plus the clinical appointment can be as high as 40 USD. This is half a month's salary for an average DRC household. As a result, patients may fail to adhere to their treatment course, increasing their chances of developing AIDS and of transmitting HIV to others. Therefore, there has been a new push to shift the role of refilling patient prescriptions to healthcare workers, nurses or pharmacists. This brings the medicine closer to the patients, and relieves the burden on clinical facilities, allowing them to devote more time to patients with advanced symptoms. In Uganda a system was trialled where patients can receive their monthly prescription from a pharmacist, then have a consultation with a nurse every three months, and a consultation with a doctor only once every six months. Both the nurses and pharmacists are trained to look for symptoms of deterioration, and will refer patients to a doctor sooner, if they feel the patient is at risk. The system reduced costs by approximately 20 per cent, with no significant difference in the health status of patients involved in the trial programme, compared to those continuing to attend a monthly clinic appointment. Although it should be noted that only stable patients (i.e. those not exhibiting symptoms of tuberculosis or other AIDS related diseases) were selected to participate in the programme.

Another system that was trialled in Mozambique put HIV patients in groups of six (Bemelmans et al., 2014). Each patient takes turns attending a clinical appointment, at which they collect the prescriptions for everyone in their group. Therefore, each patient has an appointment every six months, much more manageable than the monthly appointments they were required to attend previously. Group members are trained to look for signs of health deterioration that may require medical consultation such as excessive weight loss, early signs of tuberculosis, etc. The system has the added bonus of providing group members with a peer support base, where they can discuss issues and help each other. Children with HIV have also been included in some groups. They are taken to the appointment each month, with the relevant community member, easing the burden on their parents or legal guardian of having to take them to the clinic. They are also then given a base of people that can help them understand their medication requirements and symptoms to look for etc.

One of the main groups that fail to adhere to their medication regime are adolescents and young adults (Hudelson & Cluver, 2015). This has been attributed to difficulties in the transition from being cared for by a parent or guardian to taking responsibility for their own care. Belonging to a group throughout childhood could help them be better prepared to take charge of their own treatment as they transition to adulthood. Many of these community-based systems also include free testing and help those involved to encourage others to seek testing if they have been in a high-risk situation.

Tuberculosis

Tuberculosis is an airborne bacterial disease that generally manifests as an infection of the lungs but can affect any part of the body. Many people that carry the tuberculosis virus will never experience any symptoms. The bacteria only develop into disease for about 5–15 per cent. Some members of the community are more at risk than other including HIV positive people who are around 26 times more likely to experience TB symptoms. Although antiretroviral medication reduces the risk of developing TB, they are still twice as likely to develop symptoms than HIV negative people. Around 40 per cent of HIV related deaths are due to TB and in 2014, 12 per cent of newly diagnosed TB cases were HIV positive. Malnutrition, diabetes and smoking all also increase the risk of suffering TB symptoms.

The disease only becomes transmittable once symptoms start to occur, however, in the early stages symptoms can be very mild and similar to common cold symptoms, delaying many people from seeking medical consultation. During this period, they are highly infectious. Without treatment 45 per cent of HIV negative and nearly 100 per cent of HIV positive people who experience TB symptoms will die.

There are many medical treatments available for TB. Similar to malaria, the TB bacteria has been shown to develop resistance to the medications, however, there are at least four medications currently in common use increasing the chances of fighting off the disease. Although more recently strains of TB are being discovered that are multidrug resistant, in that they are resistant to more than one of the currently available TB medications. Therefore, continued development of TB medication is necessary to continue to effectively treat the infections.

There is also a vaccine available for TB which has helped to effectively prevent the life-threatening TB in infants and young children, however, the effects do not appear to continue into adulthood (Pai et al., 2016). Research is continuing into newer, longer-lasting vaccines.

Perhaps more important than the medical treatment is exploring the social determinants of TB related death and illness. Between 1900 and 1980 TB related deaths in western Europe and the US fell by more than 100 times, even though this predated the discovery of effective anti-TB medication (Pai et al., 2016). Improvements in hygiene and socio-economic conditions are considered to be the reason for these improvements. In fact, all three of the communicable diseases, malaria, HIV/AIDS and tuberculosis, are preventable and all have effective treatments available. Yet, they continue to persist as a leading cause of death in low and lower-middle income countries. Hargreaves et al. (2011) note that tackling the social determinants of these diseases, alongside the current regime of medical treatment will be essential to achieving the goal of eradication. They state the social determinants limit the impacts of medical treatment. These include lack of access to a medical facility, or lack of transport allowing access to a medical facility. Social stigma related to seeking a diagnosis and lack of social support during illness. They suggest provision of financial assistance, with a condition that infected household members attend medical clinic appointments for example could be an effective mechanism for supporting people experiencing symptoms to seek a diagnosis, and actively participate in a course of treatment. Adolescents and young adults with HIV/AIDS have been noted as being especially poor at adhering to course of treatment of HIV (Hudelson & Cluver, 2015). Suggested reasons for this include gender and lack of knowledge of infection, family influences, the impact of burdensome ARV treatment regimens and general attitude towards medicine, healthcare and other environmental factors, for example people living in rural areas may find it more difficult to access healthcare facilities to renew prescriptions and purchase medication, especially when they are required to take medicine several times per day. Failure to carry medication around can also lead to failures in adhering to a treatment regime, as may not have access to medicine at appropriate times.

Access to healthcare

While one aspect of the disparity between causes of death in low and high-income economies is the lack of prevention measures, the second biggest factor is the lack of access to healthcare facilities. In some cases, this can be a physical lack of access, when clinics and hospitals simply do not exist or do not have transport links. In others it is an economic lack of access where medical treatment comes at a cost that is not affordable to many of the poorer citizens. This was highlighted by Grimes et al. (2015) who investigated hospital visits in two developing economies in Africa. One hospital was located in Malawi, where healthcare is provided free of charge to all, while the other was located in Sierra Leone, where healthcare comes at a fee for everyone except children under five and pregnant or lactating women. At the hospital in Sierra Leone, approximately 12 per cent of patients absconded from the hospital after being diagnosed as requiring surgery, but before surgery was performed. The inability to pay for surgical procedures resulted in these people fleeing the hospital with serious illness that needed treatment. The researchers had no way of finding out whether or not these patients survived or received treatment elsewhere.

There are two approaches to healthcare provision in low and middle-income countries that help improve the efficiency and affordability of services. The first is removing the emphasis of consultation with a GP in a clinic as the first point of care. Similar to our discussion in the section on HIV, there are many minor ailments that could be assessed and treated by another healthcare professional, nurse or potentially even a pharmacist. In a few developing countries, this idea has been embraced, and used to improve access to health assessment for many people in an affordable way. The healthcare workers are trained to identify any symptoms that could potentially indicate something more serious, in which case they will refer the patient for consultation with a doctor.

The second approach involves taking advantage of modern telecommunication systems to allow people virtual consultation with a doctor or other medical professional, saving them the requirement to travel and wait in a clinic for an appointment, while also relieving some of the burden on medical clinics to cater to such a large number of patients. Virtual consultation allows doctors to identify probable causes of a patient's symptoms. The doctor will be able to decipher whether the patient's symptoms warrant face-to-face consultation and can advise the patient on this if necessary. Mobile phone technology can also be embraced in other ways (Chib et al., 2015). Sending patients reminders to take their medication or attend appointments. It can also be used to alert healthcare workers when they need to check up on certain patients etc. There are various examples of healthcare systems that have embraced one or both of these two approaches to providing an affordable and efficient system that is accessible to the broader population.

Medicall Home (Center for Health Market Innovations, 2018c), launched in 1998, is a one such innovation from Mexico. It is a privately funded healthcare company that charges members the equivalent of 5 USD per month. This fee allows them access to a 24-hour hotline, where a medical professional will be available to discuss their symptoms and offer advice as to whether or not consultation with a doctor will be required. The phone-line diagnosis service follows the standardised clinical protocols of the Cleveland clinic. The system has been shown to improve efficiency significantly compared to a traditional emergency service line. For example, only 14 per cent of calls that considered their situation a potential emergency ended up requiring attendance of an ambulance. The success of the system is such that a similar programme is being adopted in parts of the US.

DoctorFromHome (Center for Health Market Innovations, 2018b) is an initiative developed in India which launched in 2016. Similar to Medicall Home, DoctorFromHome uses technology to provide virtual consultation with patients who live in remote areas, and/or may find it difficult to travel. In India there are some who charge desperate people an exorbitant fee to transport them to a medical centre but have no idea whether that centre will have the correct level of healthcare available. One of the aims of the programme is to stamp out taking advantage of vulnerable people in this way. For people that have an easily identifiable illness that can be treated without face-to-face consultation, this will save them time and money. For illnesses that are identified as requiring a face-to-face consultation the doctor can advise patients on the most appropriate clinic to attend for their particular ailment.

In some regions women are disproportionately impacted by the inability to access healthcare. For example, in regions where purdah, the seclusion of women from society to limit their interaction with men, is still practised, accessing healthcare facilities can be impossible. In Pakistan, the doctHERS programme was established in 2014 to reach women in this situation (Center for Health Market Innovations, 2018a). doctHERS is entirely staffed by female physicians, and provides virtual consultations with patients, allowing a clinical consultation without the need to leave the house. They complement this service with mobile clinics staffed by female healthcare workers, to provide services for illnesses that require face to face consultation. One of the aims is to promote women in these circumstances to seek healthcare, as many often ignore their symptoms and/or don't realise that medical intervention could improve their health and prevent further deterioration of their condition.

Mental health issues can also be overlooked for poor rural communities, even though people in this situation could be more vulnerable due to the highly stressful circumstances of their life. For example, long-term unemployment and prolonged work stress have both been identified as triggers of depression along with, living in an abusive relationship, living in isolation and loneliness. Providing specialist mental healthcare can seem impossible in a situation where basic general practice is not even available. However, while treatment of mental illness is different for everyone, talking therapies can be extremely helpful for many people. Providing a community-based support system can be an effective way to prevent the onset of mental illness in vulnerable people. BasicNeeds is a mental health outreach programme founded in Kenya, but now operating in 12 countries throughout Africa and Asia (BasicNeeds, 2018). The programme provides training for community outreach workers, and even family members of those suffering from mental illness. The programme has successfully reduced hospitalisation due to mental illness in these regions and due to the evidence of its success a programme based on the model is being adopted in the US.

While the examples above are largely user pays, in Brazil, the Family Health Strategy, is a publicly funded near universal healthcare system (Macinko & Harris, 2015). The basic concept of the system originated as a maternal and child healthcare programme in the state of Ceara during the 1990s. The programme is staffed mainly by "community health agents", people with training in identifying and treating basic illnesses. Programmes will also include a doctor and nurse for referral of patients with more complex symptoms. The primary health agents are assigned approximately 150 households in their local area, which they visit once a month, regardless of whether the household members have requested medical assistance. Their role extends beyond responding to health inquiries, to also look for early signs of illness or other circumstances in households that may put residents at higher risk of requiring medical treatment, such as smoking or domestic violence. In this way they act as a force to prevent illness as much as they help treat and diagnose it. This can also assist in government interventions to promote better health including vaccination programmes. The programme has continued to expand and now in many regions incorporates other forms of medical care such as dentistry, psychology and physiotherapy.

These are just a few examples of initiatives to provide efficient and affordable healthcare to rural communities that originated in developing economies. Bhatti et al. (2017) summarise several other similar initiatives. The common thread among all these programmes is the transfer of healthcare provision closer to the community by giving better training to local healthcare professionals. This saves people the expense and time of travelling to a clinic and waiting to see a doctor for a relatively mild illness. It also increases the provision of health assessment to people who are unable or unwilling to travel when they may be experiencing symptoms of a more serious illness.

Emergency response

The death toll from traumatic injuries in emerging economies is much higher than in developed regions. While less stringent health and safety legislation is one reason for this, lack of access to emergency respondents is also a critical factor. Receiving medical attention within an hour of injury occurring can mean the difference between life and death or preventing a lifelong disability due to the injury. However, in rural areas access to emergency respondents can be virtually non-existent, especially in sparsely populated regions (see Case Study 10.1 for example). In India, although ambulance services exist in many locations, the people staffing them are essentially just drivers. They rarely receive any first aid training and when they do it is often inadequate. Thus, they cannot provide first aid to patients prior to or during transit to a medical facility, which is the most critical time for prevention of long-term disability or death. Vyas et al. (2016) instigated a training programme targeting people most likely to be first on the scene of an accident including ambulance drivers, police officers, firefighters and taxi drivers. The training programme attempted to overcome some of the limitations of previous attempts at such training. For example, many ambulance drivers may have received some form of emergency response training but were not confident about putting this into practice and admitted that they had not attempted to resuscitate a victim due to lack of confidence. The new training programme was orally based, and translated into local dialects, so that people could learn the skills even if they were illiterate or did not speak the common language. The training programme also employed the use of mannequins, allowing participants to try out the first aid techniques, rather than just reading about them which can make it difficult to feel confident putting into practice. Initial evaluation of the programme appeared successful with many participants who said they would feel comfortable attempting the first aid techniques if they came across a trauma victim.

Case Study 10.1 The Royal Flying Doctors Service (Royal Flying Doctor Service, 2017)

The difficulty of providing healthcare access to remote rural villages faces similar issues to provision of schools, in that anywhere you build a facility will only cater to a handful of people, leaving many with a long and often difficult journey. Also it will be difficult finding medical professionals willing to live in a remote community.

Australia is the sixth largest country in the world at 7,692,024 km^2 but is 54th largest population at 24.7 million. Thus, it has one of the world's lowest population densities. Issues in providing sustainable services to rural communities are exacerbated by the fact that 80 per cent of the population live within 50 km of the coast. In central Australia a population of just 3.6 million are spread out over an area of 7.13 million km^2.

Since 1928, healthcare services have been provided to this entire region by the Royal Flying Doctor Service. This provides 24-hour emergency response, regular clinics of GPs, nurses and other allied health services, and a telehealth network, that provides medical advice over the phone. The long-term success of the project shows that it is a viable and sustainable solution. Residents in remote rural areas access the same services as urban dwellers, without economically unsustainable burden of managing many healthcare facilities, for only a handful of patients, and there is no requirement for medical professionals to relocate to remote rural communities, where they do not wish to live.

Once victims arrive at the hospital having appropriately trained staff available can also be a hindrance. In Brazil for example, doctors require two years of medical training, with no specific criteria (Poggetti, 2008). Therefore, they can become fully qualified without ever having worked in emergency medicine. Doctors that do become highly skilled at emergency response surgery have generally trained at higher profile institutions in large urban areas and are unwilling to give up this lifestyle to go and work in a poor rural community.

It is not just trauma related death that is of concern. Lack of medical response can lead to long-term disability, preventing the victim from generating their income, and providing for their family. Ihekire et al. (2010) investigated hand trauma in Nigeria. Hand injuries can lead to the inability to work for many people. They identified the main cause of hand injuries as road collisions at 29 per cent, followed by work related machine operating injuries. A further 11 per cent were the result of gunshot wounds, all of which worked as security staff. The majority of these injuries were found to have been accidentally self-inflicted as security staff were armed, with inadequate training in safe handling and operation of firearms. Thus, around 40 per cent of the injuries were the result of workplace neglect. Many of these people will lose their job, as their injury will render them unable to continue performing their manual labour tasks. It will also hinder their ability to obtain another job. Holding employers accountable for staff safety and making it illegal to fire someone for being injured during work time due to an employer's neglect is essential to reduce these incidents and the subsequent burden on the economy.

Another significant cause of trauma in many emerging economies, especially those in South America, is violence. The need to promote peace and democracy in these regions will also go a long way to reducing premature deaths and disabilities.

Recommendations for best practice

Improved education is the key to better health outcomes.

Better knowledge of safe food storage and handling could prevent many food-borne illnesses.

Better understanding the mechanisms for infectious disease transmission, especially malaria, HIV and tuberculosis, could see a significant reduction in incidence of these diseases.

Better knowledge of the use of treatment for those infected with these diseases will help improve the seeking of treatment and adherence to treatment regimes. Tackling the social determinants of these diseases, alongside medical treatment will also be essential to eradicate these preventable and treatable illnesses. The provision of financial incentives could potentially be a good way to increase the willingness of young adults to seek and adhere to treatments, especially if financing allows them to develop a small business, giving them hope for the future.

An important component of increasing awareness will be the presence of healthcare professionals in remote communities. Two components to improving rural access to healthcare will be deemphasising the need for a fully qualified doctor to be the first port of call, when many diseases could be identified by a healthcare professional with basic training. The second will be embracing modern telecommunications to allow remote consultation on an electronic device. This should be sufficient to allow a doctor to determine whether a face to face consultation is necessary, in which case an appropriate clinic can be recommended to the patient.

Improved emergency response services are the final recommendation for improving health outcomes. This is especially important in areas where violence is prevalent. Increased first aid training for members of remote communities, especially those most likely to be present on the scene of a traumatic incident is recommended to reduce the number of deaths and disabilities due to trauma. Improved safety regimes are also recommended; however, these are likely to be more of a long-term goal due to the need to change people's attitudes to safety, especially employers. Implementing increased first aid training is a more immediate solution to reducing trauma related mortality.

References

BasicNeeds 2018. *Transforming lives of people experiencing mental health problems in the world's poorest communities* [Online]. Available: www.basicneeds.org/ [Accessed 19 April 2018].

Bell, M. L. & Davis, D. L. 2004. A retrospective assessment of mortality from the London smog episode of 1952: The role of influenza and pollution. *Environmental Health Perspectives*, 112, 6–8.

Bemelmans, M., Baert, S., Goemaere, E., Wilkinson, L., Vandendyck, M., Cutsem, G. V., Silva, C., Perry, S., Szumilin, E., Gerstenhaber, R., Kalenga, L., Biot, M. & Ford, N. 2014. Community-supported models of care for people on HIV treatment in sub-Saharan Africa. *Tropical Medicine and International Health*, 19, 968–977.

Benelli, G. & Beier, J. C. 2017. Current vector control challenges in the fight against malaria. *Acta Tropica*, 174, 91–96.

Bhatti, Y., Taylor, A., Harris, M., Wadge, H., Escobar, E., Prime, M., Patel, H., Carter, A. W., Parston, G., Darzi, A. W. & Udayakumar, K. 2017. Global lessons in frugal innovation to improve health care delivery in the United States. *Global Health Policy*, 36, 1912–1919.

Center for Health Market Innovations 2018a. *doctHERS* [Online]. Available: https://healthmarketinnovations.org/program/docthers [Accessed 19 April 2018].

Center for Health Market Innovations 2018b. *Doctor from home India* [Online]. Available: https://healthmarketinnovations.org/program/doctor-home-india [Accessed 19 April 2018].

Center for Health Market Innovations 2018c. *MedicallHome* [Online]. Available: https://healthmarketinnovations.org/program/medicallhome [Accessed 19 April 2018].

Chib, A., Velthoven, M. H. V. & Car, J. 2015. Health adoption in low-resource environments: A review of the use of mobile healthcare in developing countries. *Journal of Health Communication*, 20, 4–34.

Davis, F. R. 2014. *Banned: A history of pesticides and toxicology*. New Haven, Yale University Press.

Duncombe, C., Rosenbaum, S., Hellmann, N., Holmes, C., Wilkinson, L., BIOT, M., Bygone, H., Hoos, D. & Garnett, G. 2015. Reframing HIV care: Putting people at the centre of antiretroviral delivery. *Tropical Medicine and International Health*, 20, 430–447.

Grimes, C. E., Billingsley, M. L., Dare, A. J., Day, N., George, P. M., Kamara, T. B., Mkanawire, N. C., Leather, A. & Lavy, C. B. D. 2015. The demographics of patients affected by surgical disease in district hospitals in two sub-Saharan African countries: A retrospective descriptive analysis. *Springer Plus*, 4, 9.

Hargreaves, J. R., Boccia, D., Evans, C. A., Adata, M., Pettrcrew, M. & Porter, J. D. H. 2011. The social determinants of tuberculosis: From evidence to action. *American Journal of Public Health*, 101, 654–662.

Hoffman, S. L., Vekemans, J., Richie, T. L. & Duffy, P. E. 2015. The march towards malaria vaccines. *American Journal of Preventative Medicine*, 49, S319–S333.

Hudelson, C. & Cluver, L. 2015. Factors associated with adherence to antiretroviral therapy among adolescents living with HIV/AIDS in low- and middle-income countries: A systematic review. *AIDS Care*, 27, 805–816.

Ihekire, O., Salawu, S. A. I. & Opadele, T. 2010. Causes of hand injuries in a developing country. *Canadian Journal of Surgery*, 53, 161–166.

Macinko, J. & Harris, M. J. 2015. Brazil's Family Health Strategy – Delivering community-based primary care in a universal health system. *The New England Journal of Medicine*, 372, 2177–2181.

Pai, M., Behr, M. A., Dowdy, D., Dheda, K., Divangahi, M., Boehme, C. C., Ginsberg, A., Swaminathan, S., Spigelman, M., Getahun, H., Menzies, D. & Raviglione, M. 2016. Tuberculosis. *Nature Review: Disease Primers*, 2, 1–23.

Poggetti, R. S. 2008. Acute care surgeon South American model. *World Journal of Surgery*, 32, 1626–1629.

Ranson, H. & Lissenden, N. 2016. Insecticide resistance in African *Anopheles* mosquitoes: A worsening situation that needs urgent action to maintain malaria control. *Trends in Parasitology*, 32, 187–196.

Royal Flying Doctor Service (2017) About the RFDS. Available: www.flyingdoctor.org.au/about-the-rfds/ [Accessed 7 December 2017].

United Nations 2015. The millennium development goals report. New York.

Vyas, D., Hollis, M., Abraham, R., Rustagi, N., Chandra, S., Malhotra, A., Rajpurohit, V., Purohit, H. & Pal, R. 2016. Prehospital care training in a rapidly developing economy: A multi-institutional study. *Journal of Surgical Research*, 203, 22–27.

Wells, T. N. C., Huijsduijnen, R. H. V. & Voorhis, W. C. V. 2015. Malaria medicines: Glass half full? *Nature Reviews*, 14, 424–442.

World Health Organization 2015. WHO estimates of the global burden of foodborne diseases. Geneva, Switzerland: Foodborne disease burden epidemiology reference group 2007–2015.

World Health Organization 2018a. *Disease burden and mortality estimates* [Online]. Available www.who.int/healthinfo/global_burden_disease/estimates/en/index1.html [Accessed 4 April 2018].

World Health Organization 2018b. *HIV/AIDS* [Online]. Available: www.who.int/mediacentre/factsheets/fs360/en/ [Accessed 4 April 2018].

World Health Organization 2018c. *Malaria* [Online]. Available: www.who.int/mediacentre/factsheets/fs094/en/ [Accessed 4 April 2018].

11 Environment

Introduction

While it is universally agreed that the improvement of quality of life for poor rural communities should be achieved in an environmentally sustainable way, there is some discrepancy relating to what that means. A huge emphasis is placed on minimising greenhouse gas emissions (GHGs), to mitigate the impacts of climate change. However, the preservation of biodiversity is an equally important component of environmental management that is sometimes overlooked. It may seem that policies aimed at reducing GHG emissions would naturally lead to biodiversity conservation, but this is not always the case. For example, Newbold et al. (2015) modelled biodiversity impact of the four Representative Concentration Pathway scenarios (shown in Table 11.1) collected by the Intergovernmental Panel on Climate Change (IPCC). The scenario with the greatest reduction in GHG emissions (IMAGE) led to one of the highest levels of biodiversity loss, after the business as usual scenario (MESSAGE). This was due to the emphasis on biofuel use in the IMAGE scenario, which would require significant conversion of forest to cropland. There are other important environmental issues, for example depletion of the ozone layer; however, here we are focusing on those whose mitigation is likely to impact the development of rural communities, who live in and around biodiverse forest areas, as shown in Figure 11.1.

While efforts to reduce GHGs and preserve biodiversity can sometimes seem at odds, the two are in fact intrinsically linked. Alterations in climate will itself cause loss of biodiversity as temperature sensitive species may not survive. Similarly, greater biodiversity will also increase an ecosystem's resistance to climate change related impacts (Winter et al., 2017). Therefore, developing policies that maximise both these things would be mutually beneficial. Another issue facing the development of policy to protect the environment is that environmental impacts have no borders. Where one nation causes environmental damage, the impacts can affect multiple others. For example, after the Chernobyl disaster in Ukraine in 1986, radioactive waste was detected as far away as the UK, approximately 2500 km away, which resulted in restrictions on farming activities in affected areas (Andre, 2000). Therefore, to be effective, environmental policies also need to be borderless. Gaining cooperation from developing economies on

Table 11.1 Representative Concentration Pathway scenarios

Name	*Radiative forcing peak*	*Temperature increase in 2100*	*Population (billions in 2100, unless otherwise stated)*	*Description*
MESSAGE	8 W/m^2	4°C	12	Increasing crop yields and intensification account for much of the increased production required, but area of cropland and, to a lesser extent, pasture increases rapidly. Small increase in area of biofuel plantations. Urban area increases owing to increased population. Small improvements in efficiency leading to high demand for energy. Conventional oil and gas become scarce, leading to shift in favour of unconventional and carbon-intensive fossil fuels. Moderate increase in use of biofuels.
AIM	6 W/m^2	2.5°C	9.1	Urban area increases owing to human population growth. Cropland area increases to meet food demand. Pasture area declines strongly.
GCAMS	4.5 W/m^2	1.75°C	Peaks above 9 by 2065 then declines to 8.7 by 2095	Carbon pricing leads to preservation of primary forest and expansion of secondary forest. Crop yield increases, improved agricultural efficiency and dietary shifts lead to decreases in cropland and pasture areas. Small increase in area of biofuel plantations. Urban extent assumed constant. Decline in overall energy use. Decreased use of fossil fuels and increase in nuclear and renewable energy and CCS. Moderate increase in use of biofuels, but limited by availability of biomass.
IMAGE	2 W/m^2	1°C	10.1	Agriculture moves from developed to developing countries. Large increases in area of biofuel plantation. Urban extent assumed constant. Increased energy efficiency. Increased use of CCS, nuclear and renewable energy.

Adapted from Newbold et al., 2015

this scale can be challenging as often the seemingly simplest ways to bring jobs to rural areas are through environmentally damaging activities such as mining. These industries often generate significant revenue for the government itself, so it is not in their financial interest to enforce regulations that will hinder profits from these activities (Ranjan, 2017).

The global effort to manage the environment really began in 1972 with the UN Conference on Human Environment, held in Stockholm, where the United Nations Environment Programme (UNEP) was first established (Secretariat for the Convention on Biological Diversity, 2000). The next major landmark was in 1987 at the World Commission on Environment and Development (1987) (the Brundtland Commission) where the famous definition of sustainable development "development that meets the needs of the present without compromising the ability of future generations to meet their own needs" was first coined.

The 1992 the UN Earth Summit in Rio de Janeiro, Brazil, had multiple components, including the Convention on Biological Diversity. This was the first global agreement on the conservation and sustainable use of biological diversity. At this convention an agreement was signed by 150 governments, which has since expanded to 175, and sets out three goals that address:

- conservation of biodiversity;
- sustainable use of the components of biodiversity;
- sharing the benefits arising from the commercial and other utilisation of genetic resources in a fair and equitable way.

Although many countries signed the agreement, they were largely left to their own devices to establish biodiversity initiatives to address these goals and many of the initiatives were insignificant token efforts. Few led to true conservation of biodiversity.

In 2008, the United Nations launched a collaborative programme on Reducing Emissions from Deforestation and Forest Degradation in Developing Countries (REDD+) (UN-REDD, 2016). One of the aims of this programme is to ensure meaningful involvement of indigenous people and other forest dependent communities in conservation efforts. Earlier GHG reduction programmes led to concerns over the rights of these communities, as high emitting developed nations were procuring large areas of forest for preservation to offset their high emitting lifestyles, while avoiding reducing their consumptive habits. This meant local people, who were already living in poverty, had even fewer options for generating income. Forest dwelling communities often have no legal claim over the land they occupy, leaving them vulnerable to outside influences. Similarly, as described above, GHG reduction efforts do not always align with biodiversity conservation. Therefore, the REDD+ programme incorporates seven precautions that should be promoted and supported when undertaking REDD+ activities (Wallbott & Rosendal, 2018) in an attempt to avoid these two problems. These seven precautions suggest:

1 That actions complement or are consistent with the objectives of national forest programmes and relevant international conventions and agreements.
2 Transparent and effective national forest governance structures, taking into account national legislation and sovereignty.
3 Respect for the knowledge and rights of indigenous peoples and members of local communities, by taking into account relevant international obligations, national circumstances and laws, and noting that the UN General Assembly has adopted the UN Declaration on the Rights of Indigenous Peoples.
4 The full and effective participation of relevant stakeholders, in particular indigenous peoples and local communities.
5 That actions are consistent with the conservation of natural forests and biological diversity, ensuring that they are not used for the conversion of natural forests but are instead used to incentivise the protection and conservation of natural forests and their ecosystem services and to enhance other social and environmental benefits.
6 Actions to address the risks of reversals.
7 Actions to reduce displacement of emissions.

In 2010, the Convention on Biodiversity secretariat in cooperation with UNEP (meeting held in Nagoya, Japan) developed a strategic plan for protection of biodiversity with five major goals (Wallbott & Rosendal, 2018, Winter et al., 2017):

1 Address the underlying causes of biodiversity loss by mainstreaming biodiversity across government and society.
2 Reduce the direct pressures on biodiversity and promote sustainable use.
3 To improve the status of biodiversity by safeguarding ecosystems, species and genetic diversity.
4 Enhance the benefits to all from biodiversity and ecosystem services.
5 Enhance implementation through participatory planning, knowledge management and capacity building.

The strategic plan also identified five main drivers of biodiversity loss:

1 Habitat loss and degradation
2 Climate change
3 Pollution and nutrient load
4 Overexploitation and unsustainable use
5 Invasive alien species.

In 2012, at the UN Conference on Sustainable Development, Rio de Janeiro (Rio+20), the concept of a "The Green Economy" was introduced. Green economy is defined as providing:

> improved human well-being and social equity, while significantly reducing environmental risks and ecological scarcities . . . In a green economy, growth in income and employment are driven by public and private investments that reduce carbon emissions and pollution, enhance energy and resource efficiency and prevent the loss of biodiversity and ecosystem services.
>
> (United Nations Environment Programme, 2011)

Prior to the concept of a "Green Economy" economic growth and biodiversity were seen to be at odds. The storyline is that preservation of biodiversity will halt economic growth, while economic growth can only occur at the expense of conserving biodiversity. Now for the first time the potential for the two to be mutually beneficial has been acknowledged (Wallbott & Rosendal, 2018). Ultimately, the survival of humans and human society will rely on the extent to which we can preserve and increase biodiversity globally. E. O. Wilson's half-earth biodiversity sanctuary idea is one such approach, as discussed in his recent book *Half-Earth: Our Planet's Fight for Life* (Wilson, 2016).

Another issue in tackling sustainable development is defining what a good quality of life really means. Looking to high income countries as the standard for good quality living makes it near impossible to imagine how this could be achieved across the rural/urban divide without further destroying the environment. However, achieving high incomes, which allow people to have high levels of personal comfort, is not necessarily the best outcome. The 2011 documentary film *Happy* showed that the highest levels of life satisfaction are achieved by people on moderate incomes (Belic, 2011). Access to enough food, and never going hungry, and having climate appropriate shelter, where one is never too hot or too cold, are essential to a person's quality of life. This study found that having good social relationships was far more important than material wealth to many people. In fact, beyond a middle-income level the rate of happiness actually declines in wealth. Another example of the flaw of looking to higher income countries as aspiration is the diet. As China has accrued wealth dramatically in recent years, instead of seeing widespread attainment of good health through better nutrition, they have instead seen a shift towards a diet higher in animal protein and processed foods, which is leading to increases in obesity and heart disease (Dwivedi et al., 2017).

Therefore, looking to countries that are rated highly in terms of well-being and life expectancy, rather than countries with high per capita incomes could provide a better way forward. Countries that achieve this balance are known as "Goldemberg's corner", which are defined as countries that have per capita GHG emissions below 3.5 t and life expectancy greater than 70 years. Lamb (2016) takes this further by defining quality of life through life expectancy, GDP and human development index indicators, and GHG mitigation incorporating energy consumption, territorial emissions and consumption-based emissions – i.e. accounting for emissions embedded in imported products, not just those emitted locally. Countries that have achieved this improvement in development, without excessive GHG emissions, include Costa Rica (see Case Study 11.1) and Uruguay.

Case Study 11.1 Costa Rica

Costa Rica is perhaps the most famous example of a country that has achieved significant improvements in quality of life without creating excessive environmental damage. The Happy Planet Index measures how well nations are doing at achieving long, happy, sustainable lives, roughly based on the formula (New Economics Foundation, 2016):

$$\textit{Happy Planet index} \approx \frac{(\textit{Life expectancy} \times \textit{Experienced wellbeing}) \times \textit{Inequality of outcomes}}{\textit{Ecological Footprint}}$$

Costa Rica has been ranked number one three times, despite its relatively modest GDP and per capita income. Table 11.2 shows some statistics about Costa Rica. Residents of Costa Rica rank higher than the US for life expectancy and exceed both the US and the UK for well-being ratings (New Economics Foundation, 2017).

We already saw in Chapter 6 that Costa Rica has achieved 99 per cent electricity coverage, with close to 100 per cent of generation coming from renewable resources.

Costa Rica's land area is approximately 50 per cent forest and its forest cover is increasing. The increase is due to a policy of compensation to landowners for three types of forest preservation: natural forest conservation, reforestation and agroforestry. Forests contribute approximately 2 per cent of Costa Rica's GDP directly. It is also one of the 20 most biodiverse countries in the world according to the National Institute for Biodiversity and contains the ninth most biodiverse forest in the world. 27 per cent of the country is protected, and around 22 per cent of the forest area is located in protected reserves or national parks. While 19 per cent contain wildlife refuges. Costa Rica aims to be carbon neutral by 2021 and is on track to achieve the CBD 2020 biodiversity targets.

Table 11.2 Statistics on Costa Rica (New Economics Foundation, 2017)

Happiness index	*1*
GDP	$10,000
Population	4.7 million
Life expectancy	79.1 years
Well-being	7.3/10
Ecological footprint	2.8 gha/p
Inequality	15%

In this chapter we will open with a more detailed discussion on what biodiversity is and why it is important. We will then look at habitat restoration and options for preserving biodiversity in the earth's remaining forest areas. We will then discuss the concept of permaculture, i.e. ways to practise agriculture that work with nature to promote biodiversity. We then discuss how energy can be provided in ways that reduce GHG emissions and promote biodiversity (also see Chapter 6). Finally, principles underlying environmentally sustainable development are discussed in terms of energy and water, waste management, materials selection and use, ecology and how developments fit within their environmental and social contexts.

What is biodiversity and why does it matter?

The number of distinct species that are living on earth are commonly referred to as biological diversity or biodiversity. Biodiversity is impacted by various human actions, including habitat loss, conversion and degradation, habitat fragmentation, climate change, harvesting and pollution. There are two broad ideals behind the need to preserve biodiversity. Firstly, that all species inherently have the right to exist. This concept aligns with the ideals of some indigenous communities whose cultural and spiritual beliefs acknowledge the need for species preservation. For example, Choy (2018) conducted a series of interviews with a forest dwelling community in Malaysia. Each interviewee was asked how much money they would accept in exchange for giving up their forest land. Similarly, they were asked what infrastructure and services they would prefer to have access to. All of the participants gave the same response that no amount of money could replace the forest, and they could not imagine anything giving them as much pleasure as spending time in the forest landscape. They felt a sense of responsibility to future generations of their tribe to be allowed to experience the unspoilt beauty of the forest landscape which overrode any desire to have greater wealth, or better infrastructure.

Figure 11.1 Kogi village in the forest in the Sierra Nevada de Santa Marta in Colombia

On the other hand, there are members of rural communities looking to improve their quality of life through increased income generation, and agriculture as we have seen in previous chapters, appears to be the main option for many rural people. The imposition of outside regulatory bodies on preserving forest areas, without any consultation of the local people, will not stop forest destruction but lead to illegal activity, which could be even more detrimental to biodiversity preservation than controlled conversion of land to agriculture.

The inherent right of species to exist can be difficult to translate into economic value, so that it can be presented within a political landscape where decision-making often revolves around cost-benefit analyses. Hansjürgens et al. (2017) describe economic theories that incorporate social values of nature; theories that are complex and difficult for non-economic theorists to comprehend.

The second view relates to biodiversity and our need to preserve it for the ecosystem services it provides. Ecosystem services are defined as the benefits that humans derive directly and indirectly from ecosystems, which contribute manifold to human well-being. In the early ecosystem services discourse, biodiversity was not conceptualised as an ecosystem service, but as the basis of ecosystem services. However, the role of biodiversity for the provision of ecosystem services, and as an extent its contribution to human well-being is much more complicated. Ecosystem services include three types of functions essential to our continued survival (Winter et al., 2017), including:

- a support function, which includes things like nutrients or soil;
- a provisioning function, which includes food, fresh water, building materials, fuels for energy generation;
- a regulating function, such as climate regulation or water purification, O_2 respired from flora that we require to breathe.

Thus, we continually rely on the biological world and the resources it provides for survival. The richer the diversity of life, the greater the opportunity for medical discoveries, economic development, and adaptive responses to the most pressing challenges, particularly climate change.

From an economic perspective, biological resources account for at least 40 per cent of the world's economy and 80 per cent of the needs of the poor (Secretariat for the Convention on Biological Diversity, 2000). This way of thinking presents an easier way to account for the importance of biodiversity conservation from a cost-benefit analysis perspective, making it easier to present to an audience that fails to see the inherent right of other species to exist above their own needs and wants. The risk with presenting biodiversity issues in this way, is that people may become fixated on preserving the components of the ecosystem that provide services beneficial to their own needs, failing to see the value in preserving the broader spectrum of biodiversity to the benefit of the common good. While there are some components of nature we do not consume directly, ecosystems work synergistically, and every species has a role to play in that system. The removal of one part will negatively affect the whole system. For example, in the north west

US, wolves were considered an unnecessary nuisance as they posed a threat to livestock as agriculture spread into the region. Hunting of wolves was actively encouraged and during the 1920s the species was extirpated from the region. Over the following decades, significant changes to the ecosystem occurred, generally interpreted to be the result of the lack of a predator leading to increased herbivore damage to flora. Some of the most notable were the impacts on the waterways. Riparian vegetation dwindled as herbivores grazed excessively close to waterways (Beschta & Ripple, 2012). The reduced density of vegetation led to erosion of riverbanks, subsequently leading to widening of rivers in these regions. It also increased the presence of silt in the water, which hindered the ability of insects to breed. These insects provided food for birds and fish species, leading to reductions in numbers of both of these. Thus, the loss of the seemingly nuisance wolf from the region led to a decrease in riparian vegetation, leading to alteration in river flows, and a subsequent reduction in abundance and richness of species including fish, birds, butterflies, lizards and beaver. The loss of beavers in the area had a significant effect on the river, as it also meant the loss of dams.

Ecosystem impacts due to the loss of large species can be easy to see, at least in hindsight. There is, however, a lack of awareness of the importance of many smaller species. In fact, Titley et al. (2017) investigated the prevalence of species type in biodiversity studies and found disproportionate preference for studying larger species. Mammals were the focus of 12 per cent of biodiversity studies found on "Web of Science" (2016) while they account for only 0.4 per cent of known species, and birds were the focus of 13 per cent of studies, despite only accounting for 0.7 per cent of known species. On the other hand, insects which account for 70 per cent of known species (but potentially a much larger proportion of unknown) were the subject of only 23 per cent of the studies found. However, the loss of a small insect can also have a devastating impact on the bigger ecosystem. One of the key functions provided by insects is pollination, the act of transferring pollen from the anther of a flower (considered the male part) to the stigma (the female part), allowing the production of seeds and thus the continued propagation of the plant species. While there are self-pollinating plants, cross pollination is generally advantageous for avoiding the loss of genetic diversity (similar to the consequences of inbreeding in animals). Insects collect pollen from plants for their own consumption but drop pollen travelling between plants spreading the pollen. The use of pesticides that kill insects indiscriminately, such as organophosphates, has seen a decline in pollinating insect populations in some regions leading to concerns about the impacts on the broader ecosystem and agriculture.

The loss of bees is of concern as various plant/insect combinations have co-evolved and now rely on each other for continued survival. The impact of organophosphate pesticides on bees has been shown to severely compromise cognition, foraging, navigation, homing and memory abilities (Santos et al., 2016). Santos et al (2016) investigated the impacts of organophosphates on *Plebeia droryana*, an important stingless bee species found in Brazil. They exposed 441 larvae to the pesticide, and only 149 survived. Of those that survived there was an unusually high prevalence for the females to develop into sterile worker bees rather than queen bees. The main difference between the development of female bees into fertile queens is greater consumption of food during early growth, however, this process was shown to be hindered by the presence of organophosphates in the food. This lack of fertile queen bees will lead to a reduction in bee populations. The stingless bees found in Brazil are essential to the growth of many crops that form part of the Brazilian GDP, so this result is concerning from an economic perspective as well as an environmental one. Another threat to bee populations is the loss of wildflowers, due to the increased expansion of agricultural land (Papanikolaou et al., 2017). The wildflowers attract the pollinators to the area, which can spread pollination not just amongst wildflowers, but also nearby crops. Therefore, it is in the interests of farming communities to support pollinator species such as bees by using targeted pesticides and cultivating the growth of flowers alongside their crops.

Can biodiversity loss be reversed?

Many developed nations are already beginning to face the consequences of biodiversity loss. This has driven a movement to attempt restoration of biodiversity, with various approaches depending on local availability of species. In some regions, a species that was once part of the natural environment may have been extirpated from the area, but still exist elsewhere. In this case the species can be reintroduced to the region in a controlled way. In other cases, a species may have become extinct, and so a substitute species may be attempted for reintroduction. While both approaches contain risks, the second more so as the introduction of non-native species can have unpredictable negative consequences, such as did the introduction of the cane toad into Australia (see Chapter 4). However, the introduction of predatory species is generally considered more appropriate than culling by human intervention, which is often seen as unethical and does not always work. Often times, the culled species is not edible, or is not popularly eaten, and must be disposed of in landfills or other environmentally compromising way.

Take Australia, for example, where a two-fold situation of human interference has led to an unsustainable ecosystem (Lorimer et al., 2015). European settlers introduced rabbits and foxes into the region for their own amusement, while at the same time modifying and destroying the habitat of native predators, such as the dingo. Numbers of rabbits and foxes have grown out of control. Dingoes and other predators are restricted by fencing to prevent them from accessing farmland with livestock. However, it has been proposed that removing this fencing will allow predators to naturally control non-native pest species, such as rabbits, to improve the sustainability of the landscape.

One of the most famous attempts at biodiversity restoration was the reintroduction of the grey wolf (*Canus lupis*) (see Figure 11.2) into the Yellowstone National Park in north western US. As explained earlier, the wolf was extirpated from this area in the 1920s. By the 1970s the impacts of a lack of predator on the region were being noted, in particular the decline in the abundance of certain tree species that were grazed by the elk. A campaign to reintroduce the wolf into

Figure 11.2 The grey wolf in a national park

the area to help control the elk population began. In 1995/96 the reintroduction occurred. From an academic perspective this provides an excellent case study as the national park status of the area means that it is free from many outside influences, such as agricultural activities and hunting. The park has been monitored almost constantly since the reintroduction, to make careful observation of any changes in the abundance of tree species known to be grazed on by elk (Beschta & Ripple, 2012, Beschta & Ripple, 2016, Painter et al., 2015). Early on, improvements in tree species abundance were slower than predicted, and some authors asserted that these changes were due to climate change impacts rather than the presence of the wolf. However, as studies continued, the relation between climate and plant species regeneration was carefully documented alongside changes in elk presence and behavioural patterns. After a while, the changes in climate were found not to be significant enough to have caused the increased abundance of trees, and it became clearer that the presence of the wolf was at least partially responsible.

The unexpected delay between the reintroduction of the wolf and ecosystem response may be explained through the complex relationship between predator and prey. While controlling numbers of prey is often used as a measurement of predator influence, they have also been shown to significantly alter the behaviour of prey (Davies et al., 2016). Behaviours that are influenced by predators include changes in group size, vigilance, movement patterns, timing of activity and spatial distribution. Another interesting study monitored the reintroduction of lions into the Nyathi section of the Addo Elephant National Park in South Africa (Figure 11.3 shows an elephant). The park was monitored for four years following the introduction of three lions into the area. The area had not had lion presence in 100 years, and thus the existing prey had lost their evolutionary response to the presence of predators that would have been inherent in their ancestors' behaviour. The study did not find that four years was sufficient time for prey to develop a response to the presence of predators. A similar study at the Mpala Research Centre in Kenya (Riginos, 2015)

Figure 11.3 Forest elephant in Chang Rai, Thailand

found that while prey species generally displayed cautionary behaviour by eating in areas with high visibility, during periods of drought they began to graze in areas of poor visibility, where food sources were more abundant, even though they were at higher risk of attack by predators. This shows that their desire for food overrode their fear of predators, indicating some of the complexities in understanding animal behaviour and its relationship to forest management.

One of the main limitations of measuring the level of ecosystem restoration through the reintroduction of a predator is the lack of monitoring studies available from the period prior to extirpation. It is impossible to know whether the area is being restored or transformed into a new ecosystem (Lorimer et al., 2015). Painter et al. (2015) monitored aspen tree populations in Yellowstone in 1997/98 and then again in 2012. They found that aspen populations had shown significant recovery, particularly in riparian areas, and noted that while this was partly attributable to the presence of the wolves, other notable changes had occurred in the park and its surrounding area during the 15-year time period, such as an increased bison population, increased bear predation, and shifting patterns of human land use. Thus, it is unlikely that an ecosystem could ever truly be restored through the reintroduction of an extirpated species. Especially as the climate is changing and this is also having a notable effect on the behaviour of some species (Riginos, 2015). Continued monitoring of ecosystems with reintroduced species will be necessary to improve our understanding of their influence, however, the current research does suggest that the altered ecosystem is an improvement on the one that existed during the absence of the predatory species.

Reforestation

Reforestation projects can also preserve or restore biodiversity in areas that have been decimated by human activity. Case Study 11.2 shows how the actions of a single person can lead to a huge recovery in forest area. However, many reforestation projects are driven by a desire to increase carbon stores and limit the impacts of climate change. This motivation led early reforestation projects, such as those in China and South Korea, to plant cheap, easy to propagate and fast growing non-native species, such as eucalyptus, rather than many of the native tree species found in these regions (Silva et al., 2017). The easy growth of these non-native species means they could propagate and become invasive in areas of native flora, potentially decreasing biodiversity.

Case Study 11.2 Majuli Tree guy (McMaster, 2014)

The island of Majuli, located in the Brahmaputra River, Assam, India, is the largest river island in the world. Every year during the monsoon season the river floods and Majuli Island is one of the worst affected areas. Since 1917 the frequent flooding has caused erosion that has reduced the land mass by 50 per cent. The rate of erosion is increasing with time and some scientists predict the island could be eradicated in as little as 15 to 20 years. Jadav Peyang, a resident of Majuli, began planting trees on one of the most desolate parts of the island in 1979 and has continued to do so nearly every day since then. As a result he has created a dense forest covering approximately 550 hectares. The density of the forest helps protect the soil from erosion, and the root system of the plants loosens the soil structure allowing it to absorb more water during heavy rainfall reducing the impacts of flooding. Over time Jadav has seen wildlife return to the area. Up to 115 elephants have been noted, as well as rhinoceroses, deer and tigers. In 2012, vultures where spotted in the region for the first time in many years.

The improvement to the area from the tree planting project has inspired hope that similar projects throughout the Brahmaputra catchment area could help reduce devastating effects of the constant flooding suffered in the region. Jadav Peyang was given the honorary title of "The forest man of India" by the former president A. P. J. Abdul Kalam and has received many other awards and honours for his efforts. However, his forest is not safe and is constantly under threat from those who want to cut down the trees to sell for financial gain, leaving the animals that have returned to the region vulnerable. He is very angry by this prospect stating "Cutting the trees will get you nothing. Cut me before you cut my trees!" The animals themselves are under constant threat from poachers. He uses the fame he has received as an opportunity to promote ways to make Majuli more sustainable, yet none of these have been implemented. For example, he has promoted the idea of establishing a coconut tree plantation on the island. The structure of coconut trees would help prevent erosion if planted densely enough, while

(continued)

(continued)

allowing income generation through harvesting the fruit. He estimates that a plantation would take five years to become established. Local governments have applied to UNESCO to grant Majuli Island world heritage status, but the applications have been rejected.

Despite these difficulties, Jadav perseveres in his tree planting efforts, envisioning a future where the forest has been restored on most of Majuli and other areas surrounding the river. He intends to continue planting trees until the day he dies.

Brazil has committed to implementing significant reforestation as part of the Paris Climate Agreement of 2015 (Silva et al., 2017). Due to Brazil's status as home to one of the most biodiverse ecosystems in the world, ensuring the reforestation uses appropriate native species is a priority. Plant seeds for the restoration project are obtained from local nurseries. Silva et al. (2017) explored whether the supply of native seed stocks would be sufficient to fulfil Brazil's commitment to reforestation and they found that some areas would struggle to find sufficient seed stocks and expressed concerns about restoration projects lacking sufficient diversity. Collecting seeds from forests and selling to nurseries can provide an income generating opportunity to be encouraged. Hindrances to this practice, such as current legislation that restricts seed collection and a lack of trained labour capable of identifying the appropriate seeds for collection, need to be rethought for that restoration project to become successful.

Inclusion of biodiversity in life-cycle assessment

An important way to increase awareness of the importance of biodiversity is to expand the inclusion of biodiversity loss indicators to life-cycle assessment (LCA) models. LCA models are currently the most common method for incorporating environmental factors into decision-making processes for both industry and policy makers. While many LCA models exist, and some have aimed to incorporate biodiversity indicators, there is a lack of universal agreement on the standard for these, as exists for other environmental indicators such as climate change (Winter et al., 2017).

Indicators of biodiversity loss include habitat loss and degradation; climate change impacts; pollution and nutrient loss; overexploitation and unsustainable land use and invasive alien species. Biodiversity consists of three components – species diversity, ecosystem diversity and genetic diversity. Of the examples of LCA reviewed genetic diversity indicators in some were lacking. There are also many other stresses on biodiversity caused by development, such as noise and odour nuisance, which are not addressed in LCA. Another notable omission in many LCA models is diversity of assessment in relation to biotic resources (Crenna et al., 2018).

The reason for this seems obvious, biotic resources are renewable and therefore do not need to be accounted for in the same way as abiotic resources. However, their renewability is variable. For example, while all LCAs include wood, and may even specify type, none consider whether the wood was obtained from a forest area or a dedicated plantation even though each of these processes has vastly different environmental impacts. Crenna et al. (2018) suggest that establishing a renewability rate for biotic resources could help account for these more appropriately. For animals this could be their population doubling rate, while for plants it could be their rotation period. This article also suggests other factors as being important, such as the notion of resilience – the ability of a natural resource to recover after extraction by human interference, vulnerability and so on. Overall, incorporating biodiversity indicators into LCA will be important to help bring the issue onto the same footing as GHG emissions in the eyes of industry and policy makers.

Forest preservation

While the reforestation projects discussed above have met with some success in restoring local biodiversity, this process is typically limited to species that have been displaced. We cannot bring back extinct species. The Bengal tiger as shown in Figure 11.4 was nearly lost to extinction but came back from the brink due to conservation efforts in the 1970s. However, more recently it has once again come under threat due to increased poaching and the impacts of climate change threatening its habitat (World Wildlife Fund, 2018). Every time a species becomes extinct we lose a potential resource and threaten the functionality of entire ecosystems while also missing out on the services they provide us. Therefore, preserving existing forests still needs to be a priority in our efforts towards biodiversity conservation. Approximately one third of the earth's original forest cover has already been converted to agricultural land. Tropical forests are especially important as they support approximately 50 per cent of the earth's terrestrial flora and fauna, while simultaneously storing approximately 50 per cent of the world's above ground carbon (Jolin, 2018).

Figure 11.4 Forest conservation is wildlife conservation: Bengal tiger, north east India

While the remaining tropical forests make up around 10 per cent of the earth's surface, they are estimated to support around 1.5 billion livelihoods (Jolin, 2018). The majority of these tropical forests are located in developing economies and are home to very poor, uneducated communities. These people need to generate an income to not just sustain but improve their quality of life in line with the UN Sustainable Development Goals. The conversion of land to agricultural practices seems like the only way to achieve this for many of these communities. As a compromise, the concept of fracturing or fragmentation has been proposed, where part of the forest area is marked for preservation, freeing up the remainder for rural communities to develop. The impacts of fracturing are an important area of current research, and include biodiversity loss, carbon emissions and tree response studies. However, all of these areas are intrinsically linked.

There are various arguments for and against the practice of habitat fracturing. While global biodiversity is often reported as total loss of species, the functionality of ecosystems and the services they provide have been shown to be more dependent on measuring local biodiversity. An assessment of the impacts of various human activities on local diversity found that approaches to agricultural activity that minimise climate change impacts do not necessarily correspond to minimisation of biodiversity loss (Newbold et al., 2015). When establishing a fractured system, understanding and incorporating connectivity of habitats is an important component (Boscolo et al., 2017). There are three types of connectivity:

1. Structural connectivity indicates part of the landscape that is actually connected through corridors for example.
2. Functional connectivity includes species specific aspects and their interaction with landscape structures.
3. Landscape connectivity is the degree to which the landscape facilitates or impedes movement between resource patches.

Developing an understanding of the pathways that different species within an ecosystem use to connect to their food sources etc. will be necessary to allow incorporation of the most appropriate type of connectivity for a fracturing project. See Forman (1995) for an excellent discussion of the principles of landscape ecology, including the importance of patches, corridors, edges and matrices in the structural composition of landscapes at differing scales.

Haddad et al. (2015) investigated the impacts of forest fragmentation from three perspectives:

1. reduced area, the impact of species having less space;
2. increased isolation, the impact of reducing access of species to other components of the ecosystem;
3. and increased edge, bringing human settlement closer to forests.

The third aspect was of particular concern, as Haddad et al. (2015), through the use of GIS, showed that 20 per cent of the world's remaining forest area is within 100 metres of human settlement, while an alarming 70 per cent are within one kilometre. They collated observational data that had been collected for up to 35 years on multiple continents. Some of their findings are very concerning. One of the most worrying was that negative impacts on the remaining ecosystem did not plateau even in the longest observational period of 35 years but declines in species populations continued throughout the entire monitoring period. In some instances, impacts were not seen for many years after fracturing occurred. Therefore, it is difficult to tell what the true long-term impacts of fracturing could be on biodiversity in the remaining fragments.

Another threat facing some of the world's remaining forests are their location in areas that are currently, or recently have been conflict affected. Baptiste et al. (2017) explored ways forward for Colombia. Colombia is home to 10 per cent of the world's forests, and has had civil war going on from the 1960s until a peace agreement was reached in 2016. Concerns were expressed about the fate of the precious forest areas, as the country starts to ramp up development as part of the peacebuilding process. The concerns stem from previous experiences, such as the region of San Martin, Peru, where post-conflict coffee and cacao plantations were developed along with the introduction of non-native grass species. These new plantations deteriorated soil quality, making it far less absorbent leading to flash flooding and landslides. The article suggests the way forward for Colombia could avoid this by establishing forest protection legislation for a percentage of the forest area. Where agriculture is adopted, the principles of permaculture, or agri-silvi forestry, where multiple complementary crops are planted in harmony with other plants, and corridors of forest are preserved allowing species transport.

Permaculture

Permaculture is a concept developed in Australia where agriculture is practised in a biodiverse natural landscape. Harland (2017) document five examples of successful permaculture projects including the Chikukwa Project in Zimbabwe. This project was coordinated by a population of 7000 people distributed among five villages. Previous agricultural activities in the area had left the soil devoid of nutrients and compacted, so that when it rained water would run off, rather than being absorbed. After consultation with a permaculture expert they began to alter the landscape in a way that would retain the water. Simultaneously, they began composting animal waste and mixing this with the soil to restore nutrition. After three years the natural springs in the area were once again filled with water and yields from crops were increasing. Planting native trees in the area surrounding the crop fields and restricting regions of livestock grazing helped them to restore the area to a green thriving agricultural system. The community have improved their income generation, which has in turn allowed children to return to education. Therefore, development has been achieved while upholding the ideals of sustainability.

Environmentally friendly energy?

Another way, in which the triple goal of biodiversity conservation, GHG mitigation, and improving livelihoods in rural communities can lead to a conundrum, is energy provision. As we saw in Chapter 6, access to energy is a basic human right, and essential to a good quality of life. The provision of energy from renewable sources significantly reduces GHGs compared to fossil fuel burning methods. However, renewable energy systems can pose a threat to biodiversity if they are not carefully designed and managed. This issue is often overlooked but must become a part of the design process of renewable energy projects. Many of these issues are dependent on renewable energy type and these are summarised in Table 11.3. There are some issues common to all energy systems though, such as the necessity to develop infrastructure, such as access roads.

While Table 11.3 shows there are threats to biodiversity from all types of renewable energy, it also shows that these impacts can be mitigated by careful selection of appropriate locations. Incorporating biodiversity mitigation into the design stage of renewable energy projects and ensuring the location of renewable energy plants is optimised to avoid impacting biodiversity is the main mitigation strategy relevant to all technologies. The biodiversity impacts of renewable energy systems are well-studied, and the technologies are constantly being updated to reduce their impact on biodiversity, therefore ensuring the most recent technologies are employed for any new renewable energy system is another mitigation strategy.

Recommendations for best practice

When it comes to environmental regulation there is a huge emphasis on reducing greenhouse gas emissions. While this is extremely important, the loss of biodiversity could have equally detrimental impacts on our quality of life and so should be given equal weighting. Therefore, developing rural communities in an environmentally friendly way should incorporate strategies that both reduce GHG emissions and preserve biodiversity. An important aspect of this will be looking to countries that have achieved significant improvements in life expectancy and well-being, without significantly increasing GHG emissions, rather than aiming to emulate countries with the highest incomes.

Where biodiversity loss has already occurred the reintroduction of species has been shown to help maintain biodiversity in a relatively sustainable way. Although this has only been trialled where an extirpated species was returned to its once natural habitat. Where a species has become extinct, the introduction of a substitute species should be tested before it is carried out on a large scale.

The preservation of the remaining forests should be a priority. However, this must be sensitive to the rights of indigenous communities that reside in and around these areas. Careful fragmentation of forests may be an appropriate way to allow rural communities to develop while preserving biodiversity, however, the long-term impacts of this practice are not well understood, and further research will be essential to ensure the sustainability of this approach.

Table 11.3 Potential biodiversity impacts of renewable energy systems

Renewable energy system	*Potential impacts on biodiversity*
Solar	Depends on the scale of the system. A large concentrated solar power (CSP) system is made up of an array of mirrors which track the sun and continuously reflect its rays to heat a liquid, generating steam to operate a turbine. The space required to build a CSP is often large, and could threaten biodiversity depending on the current land use in that location. Selecting a location with low biodiversity is a simple way to mitigate this risk. The presence of a CSP has been shown to increase temperatures in the surrounding area, creating a microclimate effect that could pose a threat to temperature sensitive species occupying the region. Birds can be lured to solar power systems by the light, or as a result of insects that they feed on being present at these sites. Birds can then lose their life either through collision with the system or overheating if they fly into the concentrated light region of the system. Similarly, some insects can be attracted to solar systems, confusing them with their natural habitat. Laying their eggs here will hinder their ability to reproduce. PV panels on the other hand are often located on the tops of existing structures, and can therefore reduce land requirements of energy generation. Where they are accompanied by a green roof, they can actually create a habitat in an urban environment.
Wind	The rotating turbines of wind energy generating systems pose a risk to birds and bats through collisions with the blades. Birds are less at risk than bats as they generally evolve quite quickly to avoid flying into the turbine blades, although this can be species dependent, for example birds that migrate during dusk are more likely to collide with turbines due to reduced visibility. Methods for reducing the risk to birds and bats from wind turbines include improving the visibility, stopping rotation during known migration times, locating turbines away from known migration routes. Conversely, the presence of wind turbines has actually enhanced the safety of some land based species, such as tortoises.
Hydro	The obstruction of river flows caused by some hydropower plants can lead to habitat loss or damage, and fragmentation. Where dams are created for hydroplants, this can involve flooding large expanses of formerly dry land, killing the species residing here. The modification of river flows can be impacted both upstream and downstream from a hydroplant, and in different ways. The presence of a hydroplant can also affect water quality such as changes in sedimentation, overloading of nutrient cycles leading to eutrophication. Although in some cases these have been shown to stabilise over time. The mitigation of hydroplants is site specific, so it is difficult to generalise suggestions. However, designing plants to be more sensitive to biodiversity for example by incorporating technology that allows fish to pass by the plant unimpeded, will avoid upsetting their migration patterns, and altering the biodiversity of the river system. Regulation in relation to hydroplant design must be established and enforced to protect biodiversity from future developments.
Biofuels	The growing of dedicated crops for the production of biofuels poses the biggest threat to biodiversity from this type of energy generation system. The use of grasses as biofuel feedstock has been suggested as a more biodiversity friendly approach than monoculture crops such as soy. However, these grasses can become invasive and may impinge on surrounding natural areas. A better option is restoring contaminated sites that are unsuitable for food production due to the high risk of contamination. Growing feedstocks for biofuel systems as part of a permaculture system could also be a more sustainable option.
Ocean	Energy systems that harness power of the oceans are still relatively untried, and so their potential impacts on biodiversity are only theoretical at this stage. There seems little evidence that this type of energy will be developed anytime soon, so it will not be given further consideration.
Geothermal	There have been few studies into the impacts of geothermal energy generation plants on biodiversity. Some concerning issues are emissions of hydrogen sulphide and boric acid that have been measured from these plants, increased levels of arsenic have also been measured in areas surrounding geothermal plants. Although technology exists that can prevent emissions of these which should be incorporated into future developments.

Adapted from Gasparatos et al., 2017

Agricultural practices should be encouraged to be performed in a way that supports biodiversity by adhering to the principles of permaculture as far as possible. It should not be difficult to encourage this practice as it is a way to improve yields naturally through better water retention and soil nutrition.

The provision of energy should aim to use renewable sources with low carbon emissions; however, these should be developed in a way that is sensitive to local fauna and flora to ensure that it does not lead to the loss of biodiversity.

References

Andre, H. 2000. *Chernobyl radiation will affect UK 100 times longer than forecast* [Online]. Available: https://ratical.org/radiation/Chernobyl/060700.html [Accessed 29 March 2018].

Baptiste, B., Pinedo-Vasquez, M., Gutierrez-Velez, V. H., Andrade, G. I., Vieira, P., Estupiñán-Suárez, L. M., Londoño, M. C., Laurance, W. & Lee, T. M. 2017. Greening peace in Colombia. *Nature, Ecology and Evolution*, 1, 3.

Belic, R. (Director) 2011. *Happy*. USA.

Beschta, R. L. & Ripple, W. J. 2012. The role of large predators in maintaining riparian plant communities and river morphology. *Geomorphology*, 157–158, 88–98.

Beschta, R. L. & Ripple, W. J. 2016. Riparian vegetation recovery in Yellowstone: The first two decades after wolf reintroduction. *Biological Conservation*, 198, 93–103.

Boscolo, D., Tokumoto, P. M., Ferreira, P. A., Ribeiro, J. W. & Santos, J. S. D. 2017. Positive responses of flower visiting bees to landscape heterogeneity depend on functional connectivity levels. *Perspectives in Ecology and Conservation*, 15, 18–24.

Choy, Y. K. 2018. Cost-benefit analysis, values, wellbeing and ethics: An indigenous worldview analysis. *Ecological Economics*, 145, 1–9.

Crenna, E., Sozzo, S. & Sala, S. 2018. Natural biotic resources in LCA: Towards an impact assessment model for sustainable supply chain management. *Journal of Cleaner Production*, 172, 3669–3684.

Davies, A. B., Tambling, C. J., Kerley, G. I. H. & Asner, G. P. 2016. Limited spatial response to direct predation risk by African herbivores following predator reintroduction. *Ecology and Evolution*, 6, 5728–5748.

Dwivedi, S. L., Bueren, E. T. L. V., Ceccarelli, S., Grando, S., Upadhyaya, H. D. & Ortiz, R. 2017. Diversifying food systems in the pursuit of sustainable food production and healthy diets. *Trends in Plant Science*, 22, 842–856.

Forman, R. T. T. 1995. *Land mosaics: The ecology of landscapes and regions*. Cambridge: Cambridge University Press.

Gasparatos, A., Doll, C. N. H., Esteban, M., Ahmed, A. & Olang, T. A. 2017. Renewable energy and biodiversity: Implications for transitioning to a Green Economy. *Renewable and Sustainable Energy Reviews*, 70, 161–184.

Haddad, N. M., Brudvig, L. A., Clobert, J., Davies, K. F., Gonzalez, A., Holt, R. D., Lovejoy, T. E., Sexton, J. O., Austin, M. P., Collins, C. D., Cook, W. M., Damschen, E. I., Ewers, R. M., Foster, B. L., Jenkins, C. N., King, A. J., Laurance, W. F., Levey, D. J., Margules, C. R., Melbourne, B. A., Nicholls, A. O., Orrock, J. L., Song, D.-X. & Townshend, J. R. 2015. Habitat fragmentation and its lasting impact on Earth's ecosystems. *Science Advances*, 1, 9.

Hansjürgens, B., Schröter-Schlaack, C., Berghöfer, A. & Lienhoop, N. 2017. Justifying social values of nature: Economic reasoning beyond self-interested preferences. *Ecosystem Services*, 23, 9–17.

Harland, M. 2017. Permaculture: Tools for making women's lives more abundant. *Feminist Theology*, 25, 240–247.

Jolin, L. 2018. Adventures in life sciences. *Imperial*, 43, 46–47. Imperial College, London: Imperial College, London.

Lamb, W. F. 2016. Which countries avoid carbon-intensive development? *Journal of Cleaner Production*, 131, 523–533.

Lorimer, J., Sandom, C., Jepson, P., Doughty, C., Barua, M. & Kirby, K. J. 2015. Rewilding: Science, practice and politics. *Annual Review of Environment and Resources*, 40, 39–62.

McMaster, W. D. (Director) 2014. *Forest man*. YouTube [Online]. Available: www.youtube.com/watch?v=HkZDSqyE1do. [Accessed 15 September 2018].

Newbold, T., Hudson, L. N., Hill, S. L. L., Contu, S., Lysenko, I., Senior, R. A., Börger, L. et al. 2015. Global effects of land use on local terrestrial biodiversity. *Nature*, 520, 45–50.

New Economics Foundation 2016. *Happy Planet Index 2016 Methods paper* [Online]. New Economics Foundation. Available: https://static1.squarespace.com/static/5735c421e321402778ee0ce9/t/578dec7837c58157b929b3d6/1468918904805/Methods+paper_2016.pdf [Accessed 29 March 2018].

New Economics Foundation 2017. *Costa Rica* [Online]. New Economics Foundation. Available: http://happyplanetindex.org/countries/costa-rica [Accessed 29 March 2018].

Painter, L. E., Beschta, R. L., Larsen, E. J. & Ripple, W. J. 2015. Recovering aspen follow changing elk dynamics in Yellowstone: Evidence of a trophic cascade? *Ecology*, 96, 252–263.

Papanikolaou, A. D., Kühn, I., Freznel, M., Kuhlmann, M., Poschlod, P., Potts, S. G., Roberts, S. P. M. & Schweiger, O. 2017. Wild bee and floral diversity co-vary in response to the direct and indirect impacts of land use. *Ecosphere*, 8, 14.

Ranjan, R. 2017. Preserving ecologically sensitive hotspots under endogenous regulation and valuation constraints. *Natural Resource Modeling*, 30, 23.

Riginos, C. 2015. Climate and the landscape of fear in African savanna. *Journal of Animal Ecology*, 84, 124–133.

Santos, C. F. D., Acosta, A. L., Dorneles, A. L., Santos, P. D. S. D. & Blochtein, B. 2016. Queens become workers: Pesticides alter caste differentiation in bees. *Scientific Reports* 6, 9.

Secretariat for the Convention on Biological Diversity 2000. Sustaining life on earth: How the convention on biological diversity promotes nature and human well-being. *The Interim Secretariat for the Convention on Biological Diversity*. Montreal, Canada: Secretariat of the Convention on Biological Diversity, United Nations Environment Programme.

Silva, A. P. M. D., Schweizer, D., Marques, H. R., Teixeira, A. M. C., Santos, T. V. M. N. D., Sambuichi, R. H. R., Badari, C. G., Gaudare, U. & Brancalion, P. H. S. 2017. Can current native tree seedling production and infrastructure meet an increasing forest restoration demand in Brazil? *Restoration Ecology*, 25, 509–515.

Titley, M. A., Snaddon, J. L. & Turner, E. C. 2017. Scientific research on animal biodiversity is systematically biased towards vertebrates and temperate regions. *PLoS One*, 12, 14.

UN-REDD. 2016. *UN-REDD programme* [Online]. United Nations. Available: www.un-redd.org/ [Accessed 27 March 2018].

Wallbott, L. & Rosendal, G. K. 2018. Safeguards, standards and the science-policy interfaces of REDD+: Greening land use through forest-based mitigation in Costa Rica? *Journal of Environment and Development*, 27, 99–125.

Web of Science 2016. Available: https://apps.webofknowledge.com [Accessed 5 October 2018].

Wilson, E. O. 2016. *Half-earth: Our planet's fight for life*. New York: Liveright.

Winter, L., Lehmann, A., Finogenova, N. & Finkbeiner, M. 2017. Including biodiversity in life cycle assessment – State of the art, gaps and research needs. *Environmental Impact Assessment Review*, 67, 88–100.

World Commission on Environment and Development 1987. *Our common future*. Oxford: Oxford University Press.

World Wildlife Fund 2018. *Bengal tiger* [Online]. Washington DC. Available: www.worldwildlife.org/species/bengal-tiger [Accessed 22 July 2018].

12 Governance

Introduction

The implementation of the UN Millennium Development Goals disproportionately targeted urban areas leaving rural communities behind. The courses of action described in this book (summarised in Table 12.1) target rural communities in the hope that they will receive equal improvement to their urban counterparts in the new wave of UN Sustainable Development Goals, turning them from run-down areas of subsistence into Smart Villages. Selecting one action to implement in rural communities throughout the developing world is unlikely to have any significant impact on rural livelihoods because:

- Firstly, no single improvement will effect real change. For example, we saw in the energy chapter that while providing universal electrification increases income generating opportunities, these would not automatically happen, without educating the community to raise awareness of the potential for generating income, financing to allow small businesses to become established, and transport infrastructure to open up trade routes allowing small businesses access to markets beyond their local community.
- Secondly, every rural community is different, and it will take unique insider knowledge of the region's assets and customs to determine the most effective combination of policies. Real change will only be effected through the simultaneous introduction of a group of complementary measures that reflect the traditions and values of the local community.

The ability of a centralised state or federal government to achieve this goal is hard to imagine. Central government generally devises a single policy for all the rural communities under their jurisdiction which can be hundreds or even thousands of villages with differing cultural traditions, religious beliefs, employment opportunities and standards of living. There is also a tendency for central governments to divert more funding to urban areas because that is where they live, and they have a deeper understanding of the needs of the people. The advantage of central government is that they have the financial resources, building capacity and facilitation to implement these policies on a large scale (Orenstein & Shach-Pinsley, 2017). Unfortunately, the lack of community

Table 12.1 Summary of improvement strategies and the potential impact on rural livelihoods

Improvement strategy	*Potential impact on rural livelihoods*
Increased income	Ability to purchase sufficient food and stamp out hunger, maintain a more climate friendly home and better hygiene to improve health, more time to cultivate a happy family, time to spend in education.
Efficient agriculture	Increased food production, reduced hunger, reduced deforestation.
Better education	Improved job options, better health, reduced conflict, increased peace, better relationships.
Affordable housing	Better health, improved safety, reduced crime.
Energy access	Improved income generation, better health, increased safety.
Waste management	Better health, better environment, increased efficiency.
Water and sanitation	Better health, better environment.
Transport infrastructure	Increased access to services such as education and healthcare. Improved safety for children travelling to school, people travelling to work.
Health improvement	Better quality of life, increased income potential, increased educational attainment.
Environmental conservation	Better quality of life, income generation protection, i.e. reduced erosion, greater biodiversity, higher quality soils all protect income. Improved safety and better health.

participation in this top-down approach to governance often results in failure to achieve desired outcomes.

Every village will be home to a mix of potentially complementary businesses that smart economic policy could exploit to create a closed-loop system achieving both economic and environmental sustainability. Aguiñaga et al. (2018) for example studied the village of Higueras, Mexico, which included a goat cheese factory, lumber mill and oregano farm as well as many smallholder farms. The goat cheese factory had two main waste outputs: goat manure and whey. They proposed that the goat manure could be taken to a composting plant where it would be turned into fertiliser for the many smallholder farms in the community, increasing food production for the community and goat feed. The whey could be used to manufacture energy drinks and as fish bait which would open up the potential to establish a fishing business at the nearby lake which is currently unused for commercial purposes. This would mean the goat cheese factory would be zero waste and would generate the potential to establish several new businesses in the region without having to import resources from outside the village. Only a local level governance structure would have the inside knowledge of local businesses to develop such a specific economic policy.

Centralised environmental policies can often fail to achieve social sustainability goals. For example, when the region of Paragominas, Brazil was placed on the deforestation "red list" the government enforced a ban on deforestation through persecution (Viana et al., 2016). The region was the first in Brazil to be successfully removed from the red list. While this seems like a success story, prior to the ban on deforestation, slash and burn farming had been the main form of livelihood for many of the locals. The ban deprived them of the ability to grow their own food and hindered their ability to generate any income. They were simply put out of work and left unable to feed their families resulting in widespread outbreaks of violence in the region. Similarly, in the Philippines the government granted national park status to a large area of forest where indigenous tribes survived through slash and burn deforestation agriculture (Dressler & Roth, 2011). To provide an alternative means of generating income in the region the government sold the land to developers to establish an ecotourism industry. Unfortunately, the developers brought staff in from outside the area to work in the new development, only employing children from the local community to perform the lowest jobs earning less than minimum wage. The local community were once again deprived of the ability to grow their own food and with very limited income opportunities Attempts were made to teach them to grow perennial crops on the land they had already cultivated; however, they were unable to feed themselves from these and were not generating sufficient income to cover their basic needs (Dressler & Roth, 2011). These two examples highlight the need for governance to have a closer relationship with the community to ensure that policy implementation does not result in loss of livelihoods but rather has a positive benefit for the social and economic welfare of all members of the community.

The focus on achieving a single outcome rather than taking a holistic approach to governance hinders the ability of policy to achieve sustainable development. For example, focusing on broad environmental issues such as climate change mitigation has also failed to take account of localised issues such as protection of specific natural habitats or preservation of a particular species that could be important to a local community (Primdahl et al., 2013). In some regions a central government has enacted their commitment to climate change mitigation through reforestation or afforestation by planting non-native invasive species at the expense of local biodiversity (Silva et al., 2017). Similarly, many policies are formed with an economic focus and will only account for environmental and social objectives to comply with minimum requirements (Malvestio et al., 2018). There can sometimes be a commitment to marking off a checklist of measures which complies with minimum requirements but fails to achieve the broader spectrum of sustainable development (Bueren et al., 2014).

Decentralising responsibility for rural areas to more localised forms of governance thus seems a much more appropriate way to achieve sustainable development. The ideal of rural governance is that rural areas, like small islands, offer a unique opportunity: a form of government in which every politician can make a personal relationship with every elector, and in which formal government processes can be melded with informal community, and the mobilisation of non-cash resources.

Theoretically, a local government would be more in touch with the local people, and their customs and traditions than their state or federal counterparts. The local government may also be considered more likely to care about localised issues such as nature conservation as the land is their home and a legacy for their children.

Unfortunately, the reality is often a somewhat different picture. Firstly, in many regions decentralisation has been insincere and the local government officials simply acted as puppets for the central government to put forth their agenda and stamp out dissidents (Bebbington et al., 2006, Crook, 2003). In other regions local governance committees have been given the freedom to develop policies, but not the power and/or the resources to effectively implement them (Blatter, 2004). Perhaps the most common problem is local governance systems that have true power and resources are often formed of local elites, who abuse their position for personal gain. Good and Beatty (2011) claim this is inevitable as "Individuals acting in rational pursuit of their own self-interest, will sacrifice the long-term viability of a shared resource for short-term gain." It has been argued that local governments are even less likely to invest in policies for alleviating poverty than a centralised government because they can benefit directly from diverting funds to their own cause. Taking education as an example, a person is most likely to invest in the school their children attend ensuring they have the best possible facilities and advantages for a good start in life. They have little incentive to invest in establishing education systems for the poor.

So, what is good governance? It has been argued that "There is no universally accepted definition of governance. There is not even a consensus on which set of phenomena can be properly grouped under the title of governance", governance takes different forms across nations (Beer, 2014). Bezboruah and Pillai (2015) define governance by six dimensions as illustrated in Figure 12.1. Blair (2000) states that governance must be democratic or just and equitable. They go on to define this as meaningful authority devolved to local units of governance that are accessible and accountable to the local citizenry, who enjoy full political rights and liberty. Citizens must have a meaningful role in decision-making; and the government must be held accountable for how decisions affect them. Thus, the key elements of good governance are to ensure that citizens from all sectors of the community have a voice in the decision-making process and that elected officials operate transparently and are held accountable for their actions.

Giving citizens a meaningful voice

One method of attempting to ensure a voice is by requiring a set number of seats on the council to be filled by women, members of indigenous tribes and minorities. Blair (2000) looked at examples from Bolivia, Philippines, Mali and India. The schemes were only partially successful and fared better for indigenous tribes and minorities than it did for women. In many instances, local elites had commandeered the election to ensure the required representativeness was fulfilled, without giving a true voice to the people of these groups. For example, many of the women were wives of local elites and their contributions to the council were

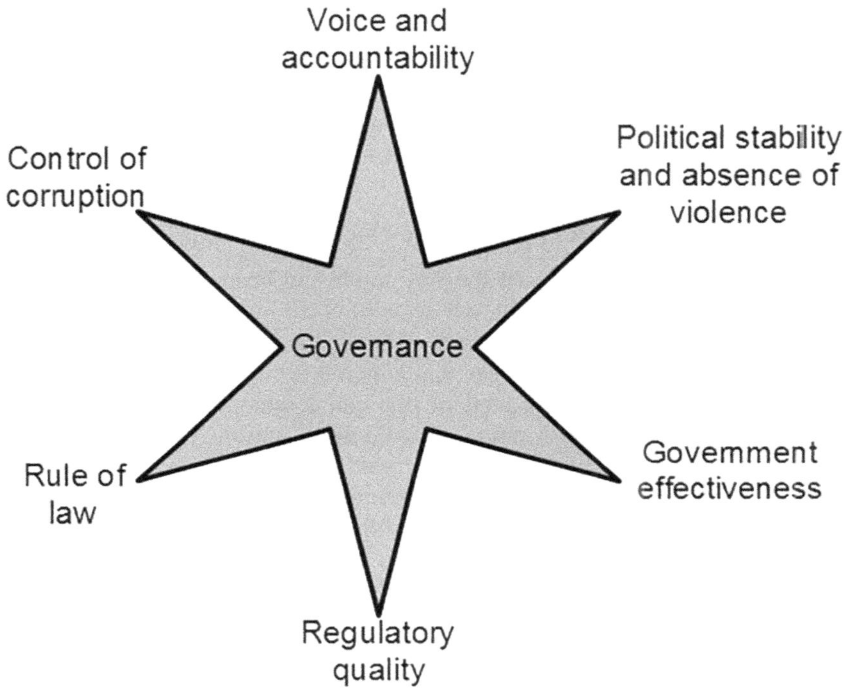

Figure 12.1 Dimensions of governance

dictated by their husbands, rather than a representation of their own opinions. Similarly, some of the representatives from indigenous tribes and minorities were employees of local elites and were contributing their boss's opinions in order to hold onto their jobs. There is also difficulty overcoming the attitudes of local elites that the poor are innately inferior. In Zimbabwe attempts to open up seats on the local council to people from poorer communities were met with strong opposition (Crook, 2003). Local elites saw the lack of education of the poorer members of the community as an indication that they would not have the intellectual capacity to comprehend policies and would therefore not be able to make a meaningful contribution to the decision-making process.

It is also important that policy does not go too far the other way. Policies that target the poorest members of the community only alienate local elites. They still have an important role to play in the local community and it is important to avoid this scenario. Blair (2000) describes the three approaches to governance as the traditional "rich get richer" scenario where the council is formed of local elites diverting funds to their own needs; "poverty-focused" where all funds are diverted to the poorer areas of the community. The ideal scenario is "universalistic" where programmes that benefit everyone are established. Taking education as an example in the first scenario existing schools continue to get more funding, leaving children outside of these missing out on education; in the second scenario funding is targeted at establishing schools in poor areas. This can be a false positive for example without the backing of local elites it will be difficult to attract teachers to the area to work in the schools. In the final scenario "primary education for all" is the goal. This will appeal to both elites and minorities and therefore is more likely to get backing. It also provides a more long-term solution.

Government community partnerships/local leadership

Pal and Wahhaj (2017) assessed the way funds had been spent in rural communities of Indonesia since decentralisation of fiscal responsibility in the early 2000s. They found a strong relationship between the way funds were spent and local culture. For example, villages that adhered to "Adat", an unwritten traditional code that covers all aspects of personal conduct, were more likely to have developed better infrastructure since decentralisation. These villages had informal leaders that had acted in an unofficial governance role prior to decentralisation. Thus, the transition for these communities was smoother because the local community already had strong leadership. However, the infrastructure was focused on internal village things. There was a distinct lack of development that improved connectivity between adjacent villages such as roads. This showed a reluctance of local leaders to allow outside influences into their area, and risk losing their hold on power.

This concept of small close-knit communities having informal leaders that people turn to in times of need is not unique to Indonesia. In remote rural areas globally, people may not trust government officials or feel that they do not have a way to communicate their concerns with the government. Exploiting the authority and influence of these informal leaders through the development of government-community partnerships can be an effective way to improve development. Beer (2014) explored two different approaches to government community partnerships in Australia. The first was a community leadership formed of local business operators. Although the group claimed they were working with the government to promote the best interests of their community, the lack of diversity on the committee was a concern for many members of the community who felt the reps were pushing their own agenda to increase profits for their businesses. Thus, the risk of local elites gaining undue influence is also of concern in this style of governance. The risk is in some ways worse because members of these informal committees are not held to the same level of accountability as government officials (Pemberton & Goodwin, 2010). The second approach Beer (2014) explored was a government appointed community task force. Concerns were raised that because the government appointed the members they were pushing their own agenda. However, the task force disagreed with government policy initiatives and worked hard to negotiate for policy changes to better represent the local community. Overall, both considered approaches had some victories and some losses, which in the eyes of the government was an overall success. However, for the local community the failures which included diminished economic circumstances and reduction in population had a more significant impact on their livelihoods than the successes making it hard for them to see the upside.

The European LEADER programme is an example of this type of community-led local development that has been around since the early 1990s. The programme is implemented by local action groups, of which there are around 2600, covering 54 per cent of the European rural population (European Network for Rural Development, 2018). One of the key elements of establishing LEADER is ensuring that all members of the community feel they are welcome to participate and have a relevant voice. In the EU's guide to LEADER it states that:

> merely informing local people that they can contribute to the development of a local development strategy and a bid for funding for their area is not enough. Local action groups have to motivate and enable real participation, plan whose involvement you seek and how you enable this. It is important to ensure that your process is as inclusive as is feasible, avoid excluding groups e.g. the young or the elderly who may face difficulties such as ability to travel or time constraints.
>
> (European Network for Rural Development, 2016)

In Poland around 47 per cent of villages have instigated LEADER, which begs the question what is stopping the rest of the villages from participating? Fałkowski (2013) hypothesised that uptake of the LEADER programme would be related to the level of security felt by local government officials. Their study found the scheme was implemented more successfully in areas with fewer council members and a majoritarian voting system. Both of these factors indicate that it is easier to hold officials accountable and thus they are more strongly motivated to instigate community partnerships. Politicians whose position in office is more secure either because there are more council members, making it more difficult to identify who is accountable for poor policy, or because there is a proportional election system, where citizens vote for a party rather than an individual person. Thus, greater accountability motivates politicians to seek greater community participation.

In Nepal, the implementation of microhydro electricity plants was managed locally by a diverse selection of community members (Legros et al., 2011). The committee ensured that electrification reached all households, even the poorest, and prevented it from being commandeered by the wealthiest community members. However, these systems are still vulnerable due to the impacts of climate change which will alter the water levels in the waterways, impacting the ability of the microhydro plants to continue functioning (Islar et al., 2017). Policies should aim to be intergenerational, providing a resilient system that will last for future generations. Case Study 12.1 also describes an example of community-led governance.

Case Study 12.1 Ringarooma Tasmania Irrigation Project (Nogrady, 2017)

Tasmania, like most of Australia, suffers periodically from droughts. To control water availability, the government has a system of sensors in the main waters that periodically inform them of the water levels. When water levels fall below a certain threshold, the government issues a ban on irrigation, until water levels are replenished. This is a justifiable approach given past consequences of poorly managed irrigation (for example see the Aral Sea case study in Chapter 4). However, the farmers in Tasmania feel that managing water at the state level is unfair, as rainfall occurs differently across the state, and one catchment does not reflect another. The government agreed to trial a self-regulation approach with the farmers of the Ringarooma catchment area. This catchment supports around 70 farmers. During the trial, the data collected by the water sensors was sent to the farmers via text message, and the farmers were allowed to decide when it was appropriate to irrigate their crops. This was a test to see whether a farming community could be trusted to "do the right thing" when it came to irrigating their crops. The results of the trial were a success, with the farmers acting responsibly to only irrigate when water levels were appropriately high. Although this could potentially lead the way to passing on responsibility for management to the users, this situation may be unique in that all the farmers worked together as a community to make decisions. In other catchment areas, farmers may be rivals, and may not be able to agree on appropriate irrigation in the same way.

Communication between governments and citizens

In the two previous sections we have described ways to ensure citizens have a voice, the other essential component to good governance is ensuring accountability of political leaders. Instruments to accountability include: elections; political parties (competition); civil society; media; public meetings; formal procedures; opinion surveys. The cornerstone for all of these instruments is having effective communication between the government and its citizens.

Barriers to government – citizen communication can take many forms. People living in a region that is new to democracy may still be wary of speaking up for fear of persecution. In regions that are transitioning from an autocratic system or coming out of war it can be particularly difficult for people to feel safe speaking out against the government. Mvukiyehe and Samii (2017) studied the UN assisted transition of several rural communities in post-war Liberia. In many of these regions local elites were glad to maintain the fear of speaking out because they did not want their authority threatened. The UN has various intervention strategies to aid citizens into a democratic society. In Liberia two approaches were tried: the first was hosting a monthly civic education programme during which UN representatives would engage with citizens about democracy and encourage them to speak out against any threats they received from politicians. The second was security presence to prevent reprisals against citizens for voting in elections. While both interventions led to positive outcomes, the civic education sessions were the most effective at encouraging citizens to participate in the democratic process. Another positive outcome was that participating in the civic education classes did not invite any backlash in the form of threats or violence.

Barriers to communication can also be physical such as the inability to attend public meetings due to time constraints for those needing to work long hours to make ends meet, or lack of transport to access the location of the meeting. The availability of media outlets can also be a hindrance. As we discussed in Chapter 6 many rural communities lack electricity and therefore would not be able to operate televisions or radios. Many areas do not have newspapers, and/or many citizens cannot read. In other regions while access to these forms of media is available they are controlled by the government and will not inform citizens of misdeeds or failures. Thus, there are many barriers to communicating with citizens who their local politicians are, what policies they are implementing and the intended outcomes of those policies. In many regions governments have been implementing policies aimed at improving livelihoods of the poor. However, in many instances this has been a top-down implementation where the government decides what is best for the people. Assessment of the effectiveness of this process has found them to be ineffectual. For example, the government of Mozambique passed new legislation to help the residents in the area surrounding the Maputo Elephant Reserve. These people rely on hunting in the reserve as both their source of food and income, and their livelihood is being threatened by people from outside hunting in the reserve, leading to stocks dwindling below a safe threshold. Soto et al. (2001) surveyed the local community to ascertain their views on the new legislation. They found that most of the residents had never heard of the legislation, and those that had, did not believe that it would directly benefit them. This story highlights the need for the government to communicate with citizens about new policies.

It is therefore important to ensure that people firstly feel there will not be repercussions for contributing to political discussions and secondly have a physical means of contributing. Thus, introducing a simpler mechanism for people to participate in governance decision-making is essential. Exploiting the abilities of modern technology has been proposed as an efficient and effective way to achieve this. While many regions lack things like transport and employment opportunities restricting their ability to attend town meetings etc. ownership of mobile phones is becoming fairly ubiquitous. It was reported that 78 per cent of the world's households had a mobile phone in 2011. Even in sub-Saharan Africa, the lowest penetration rate, the per centage is reported to be 67 per cent and rising (Source Trace, 2018). Thus, the use of mobile technology for communication between citizens and government is becoming the way of the future. Mobile apps

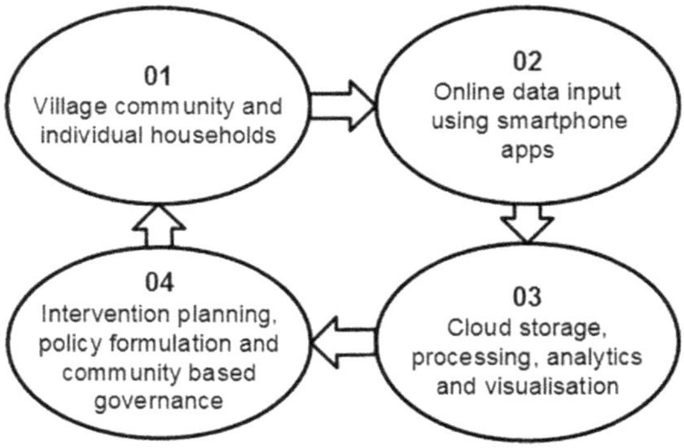

Figure 12.2 Design of community-based smart governance processes

are already being employed around the world. In South Africa mobile apps are being used to communicate with smallholder farmers informing them of things like weather conditions and soil temperature, helping them to optimise their yields by timing sowing, watering and fertilising more precisely. It is also used to inform them of commodity prices to help them decide on the best time to sell their produce at markets and when to store it (Peisker & Dalai, 2015).

As part of a research project within the Smart Villages Lab (SVL) at the University of Melbourne, Australia, the authors developed a unique Information and Communication Technology (ICT) based prototype connecting the rural community with the public administration in the state of Assam, India. Figure 12.2 shows the four-step process model of the prototype where socio-economic household data is collected for analysis and support of governance for the rural community. The prototype, named the Smart Data Platform, has been validated using data collected from over 2300 households across 37 local villages on Majuli Island, Assam. The concept facilitates rapid assessment of community socio-economic conditions and needs by analysing data from the survey and integrating it with data from external sources. Integration capability of the external data assists in analysing how various intervention efforts on the selected community are aligned to the needs and the requirements exposed though the survey data and thereby fine tunes the effort for achieving better governance outcomes.

Intergovernmental communication

Governments are comprised of many different departments each assigned responsibility for a particular sector of society. In some instances, each of these departments are working independently to develop policies that are going to impact the same citizens. These departments should be working collaboratively to develop policies that are complementary. For example, in the agriculture chapter we discussed the potential of integrated manure management for providing an energy source and offsetting the use of synthetic and mineral fertilisers in crop farming. Instead, many agricultural departments are subsidising the use of synthetic and mineral fertilisers, and manures are discarded into the environment where they are causing harm to both human health and the environment (Teenstra et al., 2014). To implement integrated manure management the agricultural department of the government would need to work together with the energy department and the waste management department and even potentially the environmental management department depending on how the government is structured. Where manure management policies exist, they often fail to achieve integration and are largely unenforced. Bueren et al. (2014) describe this as organised irresponsibility and explain that it stems from too many disconnected players in the system. The farmers and the markets they supply need to work together with the multiple levels of government meaning there is no single entity responsible for ensuring the success of the system and thus no one to hold accountable when the system fails.

Technology could also play a significant role in improving intergovernmental communication by providing different departments real time access to activities in complementary areas. For example, new policy propositions could be uploaded to a shared database and marked with alerts for any departments that may be potentially impacted by them. This way these departments can consult on the final form of the policy before it is implemented, ensuring policies from different departments complement rather than contradict each other.

The role of NGOs

There is currently a climate of corporate social responsibility (CSR) in many developing countries, along with government promises of improving quality of life for the poor. This brings into question whether NGOs are still necessary to help with the implementation of such policies. RGVN argues that the role of the NGO is not replaceable because the commitment of governments and corporations is not sincere (Personal Communication). Their true goal is to enhance their reputation and increase profit and will therefore provide token assistance without effecting real change. For example, many CSR programmes commit to three years of provisions. This random selection of a time frame for implementation of a project does not make sense, as they have no way of knowing how long it will take the community to become self-sufficient. Non-profit organisations such as RGVN commit to sustaining a project until the community is truly self-sufficient. This cannot be achieved when a random time limit is applied to the project. Corporations also do not engage with the community before deciding on what projects they are going to implement and can seem imposing and interfering to the community who have no say in what is happening in their village. RGVN on the other hand start off their projects by arranging a face to face meeting with community members to discuss options for improving their quality of life and giving them a say in how the implementation will occur.

References

Aguiñaga, E., Henriques, I., Scheel, C. & Scheel, A. 2018. Building resilience: A self-sustainable community approach to the triple bottom line. *Journal of Cleaner Production*, 173, 186–196.

Bebbington, A., Dharmawan, L., Fahmi, E. & Guggenheim, S. 2006. Local capacity, village governance, and the political economy of rural development in Indonesia. *World Development*, 34, 1958–1976.

Beer, A. 2014. Leadership and the governance of rural communities. *Journal of Rural Studies*, 34, 254–262.

Bezboruah, K. C. & Pillai, V. 2015. Exploring the participation of women in financial cooperatives and credit unions in developing countries. *Voluntas*, 26, 913–940.

Blair, H. 2000. Participation and accountability at the periphery: Democratic local governance in six countries. *World Development*, 28, 21–39.

Blatter, J. 2004. "From spaces of place" to "Spaces of flows"? Territorial and functional governance in cross-border regions in Europe and North America. *International Journal of Urban and Regional Research*, 28, 530–548.

Bueren, E. M. V., Bueren, E. T. L. V. & Zijpp, A. J. V. D. 2014. Understanding wicked problems and organized irresponsibility: Challenges for governing the sustainable intensification of chicken meat production. *Current Opinion in Environmental Sustainability*, 8, 1–14.

Crook, R. C. 2003. Decentralisation and poverty reduction in Africa: The politics of local-central relations. *Public Administration and Development*, 23, 77–88.

Dressler, W. & Roth, R. 2011. The good, the bad, and the contradictory: Neoliberal conservation governance in Rural Southeast Asia. *World Development*, 39, 851–862.

European Network for Rural Development 2016. LEADER local development strategies (LDS): Guidance on design and implementation. Brussels: European Commission.

European Network for Rural Development 2018. *LEADER/CLLD* [Online]. Available: https://enrd.ec.europa.eu/leader-clld_en [Accessed 19 June 2018].

Fałkowski, J. 2013. Political accountability and governance in rural areas: Some evidence from the Pilot Programme LEADER+ in Poland. *Journal of Rural Studies*, 32, 70–79.

Good, A. G. & Beatty, P. H. 2011. Fertilizing nature: A tragedy of excess in the commons. *PLoS Biology*, 9, 1–9.

Islar, M., Brogaard, S. & Lemberg-Pedersen, M. 2017. Feasibility of energy justice: Exploring national and local efforts for energy development in Nepal. *Energy Policy*, 105, 668–676.

Legros, G., Rijal, K. & Seyedi, B. 2011. Decentralized energy access and the millennium development goals: An analysis of the development benefits of micro-hydropower in rural Nepal. Rugby, UK: United Nations Development Program.

Malvestio, A. C., Fischer, T. B. & Montaño, M. 2018. The consideration of environmental and social issues in transport policy, plan and programme making in Brazil: A systems analysis. *Journal of Cleaner Production*, 179, 674–689.

Mvukiyehe, E. & Samii, C. 2017. Promoting democracy in fragile states: Field experimental evidence from Liberia. *World Development*, 95, 254–267.

Nogrady, B. 2017. *Tasmanian irrigators pioneer high-tech, localised water management* [Online]. CSIRO. Available: https://blogs.csiro.au/ecos/sense-t/ [Accessed 14 June 2018].

Orenstein, D. E. & Shach-Pinsley, D. 2017. A comparative framework for assessing sustainability initiatives at the regional scale. *World Development*, 98, 245–256.

Pal, S. & Wahhaj, Z. 2017. Fiscal decentralisation, local institutions and public good provision Evidence from Indonesia. *Journal of Comparative Economics*, 45, 383–409.

Peisker, A. & Dalai, S. 2015. Data analytics for rural development. *Indian Journal of Science and Technology*, 8, 50–60.

Pemberton, S. & Goodwin, M. 2010. Rethinking the changing structures of rural local government – State power, rural politics and local political strategies. *Journal of Rural Studies*, 26, 272–283.

Primdahl, J., Kristensen, L. S. & Swaffield, S. 2013. Guiding rural landscape change: Current policy approaches and potentials of landscape strategy making as a policy integrating approach. *Applied Geography*, 42, 86–94.

Silva, A. P. M. D., Schweizer, D., Marques, H. R., Teixeira, A. M. C., Santos, T. V. M. N. D., Sambuichi, R. H. R., Badari, C. G., Gaudare, U. & Brancalion, P. H. S. 2017. Can current native tree seedling production and infrastructure meet an increasing forest restoration demand in Brazil? *Restoration Ecology*, 25, 509–515.

Soto, B., Munthali, S. M. & Breen, C. 2001. Perceptions of the Forestry and Wildlife Policy by the local communities living in the Maputo Elephant Reserve, Mozambique. *Biodiversity and Conservation*, 10, 1723–1738.

Source Trace 2018. *How mobile apps are helping agriculture in achieving sustainable development* [Online]. Available: www.sourcetrace.com/mobile-apps-for-agriculture/ [Accessed 19 June 2018].

Teenstra, E., Vellinga, T., Aektasaeng, N., Amatayakul, W., Ndambi, A., Pelster, D., Germer, L., Jenet, A., Opio, C. & Andeweg, K. 2014. Global assessment of manure management policies and practices. Wageningen: Wageningen UR Livestock Research.

Viana, C., Coudel, E., Barlow, J., Ferreira, J., Gardner, T. & Perry, L. 2016. How does hybrid governance emerge? Role of the elite in building a green municipality in the eastern Brazilian Amazon. *Environmental Policy and Governance*, 26, 337–350.

Epilogue

India is one of the world's largest and fastest growing economies. It is also a place where the gap between people's quality of life in rural communities and that in cities is rapidly widening. India has been trying to eradicate poverty in rural areas through expanding development programmes aimed at meeting the needs of the rural poor. While the policies and governance arrangements needed to effectively deliver rural development programmes to achieve these goals are still evolving, many grand schemes targeting the rural poor have often not delivered the expected outcomes.

One of the key areas of disparity in India between urban and rural settlements is the provision of improved housing and infrastructure. That is why new, smart approaches to housing and infrastructure is at the core of the current book. In it the authors have tried to lay out some best practice approaches for key areas of consideration, which have mostly been drawn from literature across a range of areas relevant to development of the rural communities. While the book provides a snapshot of key focus areas, which the authors feel is essential for creating smarter villages, this is still an ongoing process. Research currently being undertaken by the Smart Villages Lab (SVL), based in the University of Melbourne's Faculty of Architecture Building and Planning, is continuing to produce knowledge needed to build capacity for creating smarter rural communities for the future. The body of research that those involved with the SVL are pursuing focuses on better understanding the needs of rural communities in the state of Assam, India. The aim of this research is to inform decision-making with respect to rural Smart Village development in India. The Smart Villages Lab, and the knowledge base it is building, however, seeks to help in creating smarter rural villages not just in India but also in other regions of the world where rural communities need improved living conditions, housing and infrastructure.

The Smart Villages Lab (SVL) is, for example, currently undertaking research exploring the housing and infrastructure needs of people living in 37 rural villages in the state of Assam, India. This research involves extensive surveying of over 2300 households within villages on Majuli Island, located in the mighty Brahmaputra River. The data being collected for that project is being plotted using a purpose built geographic information system (GIS) to study the spatial distribution of the villagers needs and linking their needs with their socio-economic characteristics. Building on this body of bottom-up derived empirical data, the SVL is continuing to reveal new understanding about what is needed to improve the living conditions of these people, developing new methodological processes and revealing theoretical insights, some of which contrast with existing scholarly viewpoints.

The authors, coming from differing academic and professional backgrounds, have, through these research efforts, come to realise several things. Firstly, the challenges presented by rural development require innovations that are not necessarily applicable to urban areas and cities. Secondly, rural communities often exhibit resilience, resourcefulness and a sense of community that is lacking in cities, but qualities which many cities desperately need. Thirdly, and most importantly, rural areas, particularly in developing countries, must seek to be made as socially, environmentally and economically sustainable as possible in the face of ever increasing environmental and social changes the planet is currently experiencing and will continue to experience into the future.

Bibliography

Abdul Razak, N., Abdul Malik, J. & Saeed, M. 2013. A development of smart village implementation plan for agriculture: A pioneer project in Malaysia. Proceedings of the 4th International Conference on Computing and Informatics, ICOCI 2013 Paper No. 28–30 August, Sarawak, Malaysia. Universiti Utara Malaysia.

Abdurahman, A. Z. A., Ali, J. K., Khedif, L. Y. B., Bohari, Z., Ahmad, J. A. & Kibat, S. A. 2016. Ecotourism product attributes and tourist attractions: UiTM Undergraduate Studies. *Procedia – Social and Behavioral Sciences*, 224, 360–367.

Abo-elazm, F. M. & Ali, S. M. 2017. The concept of "Local Smart Architecture": An approach to appropriate local sustainable buildings. *International Journal of Cultural Heritage*, 2.

Abu-Ghazzeh, T. M. 1997. Vernacular architecture education in the Islamic society of Saudi Arabia: Towards the development of an authentic contemporary built environment. *Habitat International*, 21, 229–253.

ADAS 2009. Phosphorus: A vital non-renewable resource. Durham: Premier Waste Management.

Aguiñaga, E., Henriques, I., Scheel, C. & Scheel, A. 2018. Building resilience: A self-sustainable community approach to the triple bottom line. *Journal of Cleaner Production*, 173, 186–196.

Akinyele, D. O. & Rayudu, R. K. 2016. Techno-economic and life cycle environmental performance analyses of a solar photovoltaic microgrid system for developing countries. *Energy*, 109, 160–179.

All Power Labs 2018. *How gasification works* [Online]. Berkeley, California. Available: www.allpowerlabs.com/gasification-explained [Accessed 13 March 2018].

Allison, T. H., Davis, B. C., Short, J. C. & Webb, J. W. 2015. Crowdfunding in a prosocial microlending environment: Examining the role of intrinsic versus extrinsic cues. *Entrepreneurship Theory and Practice*, 53–73.

Alveano-Aguerrebere, I., Ayvar-Campos, F. J., Farvid, M. & Lusk, A. 2017. Bicycle facilities that address safety, crime, and economic development: Perceptions from Morelia, Mexico. *International Journal of Environmental Research and Public Health*, 15, 22.

Anctil, A. & Blanc, D. L. 2016. An educational simulation tool for integrated coastal tourism development in developing countries. *Journal of Sustainable Tourism*, 24, 783–798.

Andre, H. 2000. *Chernobyl radiation will affect UK 100 times longer than forecast* [Online]. Available: https://ratical.org/radiation/Chernobyl/060700.html [Accessed 29 March 2018].

Anna-Maria, V. 2009. Evaluation of a sustainable Greek vernacular settlement and its landscape: Architectural typology and building physics. *Building and Environment*, 44, 1095–1106.

Arnold, B. 2014. *Collecting and storing energy from wind turbines* [Online]. Available: www.azocleantech.com/article.aspx?ArticleID=488 [Accessed 13 March 2018].

Asian Development Bank 2003. *Road funds and road maintenance: An Asian perspective.* Manila, Philippines: Asian Development Bank.

Asomani-Boateng, R., Fricano, R. J. & Adarkwa, F. 2015. Assessing the socio-economic impacts of rural road improvements in Ghana: A case study of Transport Sector Program Support (II). *Case Studies on Transport Policy*, 3, 355–366.

Attanasio, O., Ausburg, B., Haas, R. D., Fitzsimons, E. & Harmgart, H. 2015. The impacts of microfinance: Evidence from joint-liability lending in Mongolia. *American Economics Journal: Applied Economics*, 7, 90–122.

Aulakh, J. & Regmi, A. 2013. Post-harvest food losses estimation: Development of consistent methodology. FAO.

Australian Government 2010. The cane toad fact sheet. Canberra: Department of the Environment, Water, Heritage and the Arts.

Australian Government 2016. *The school of the air and remote learning* [Online]. Canberra Available: www.assoa.nt.edu.au/visitors-centre/the-centre/history [Accessed 15 March 2018].

Azuri Technologies Ltd 2017. *Paygo solar power* [Online]. Cambridge, UK. Available: www.azuri-technologies.com/what-we-do [Accessed 21 February 2018].

Banda, K., Sarkar, R., Gopal, S., Govindarajan, J., Harijan, B. B., Jeyakumar, M. B., Mitta, P., Sadanala, M. E., Selwyn, T., Suresh, C. R., Thomas, V. A., Devadason, P., Kumar, R., Selvapandian, D., Kang, G. & Balraj, V. 2007. Water handling, sanitation and defecation practices in rural southern India: A knowledge, attitudes and practices study. *Transactions of the Royal Society of Tropical Medicine and Hygiene*, 101, 1124–1130.

Banerjee, A., Chandrasekhar, A. G., Duflo, E. & Jackson, M. O. 2013. The diffusion of microfinance. *Science*, 341.

Banerjee, A., Duflo, E., Glennerster, R. & Kinnan, C. 2015a. The miracle of microfinance? Evidence from a randomized evaluation. *American Economics Journal: Applied Economics*, 7, 22–53.

Banerjee, A., Duflo, E., Goldberg, N., Karlan, D., Osei, R., Parienté, W., Shapiro, J., Thuysbeart, B. & Udry, C. 2015b. A multifaceted program causes lasting progress for the very poor: Evidence from six countries. *Science*, 348, 772–790.

Banerjee, N., Stearns, E., Moller, S. & Mickelson, R. A. 2017. Teacher job satisfaction and student achievement: The roles of teacher professional community and teacher collaboration in schools. *American Journal of Education*, 123, 203–241.

Banerji, R. (Director) 2015. *Voices from the forest.* India.

Bański, J. & Wesołowska, M. 2010. Transformations in housing construction in rural areas of Poland's Lublin region—Influence on the spatial settlement structure and landscape aesthetics. *Landscape and Urban Planning*, 94, 116–126.

Baptiste, B., Pinedo-Vasquez, M., Gutierrez-Velez, V. H., Andrade, G. I., Vieira, P., Estupiñán-Suárez, L. M., Londoño, M. C., Laurance, W. & Lee, T. M. 2017. Greening peace in Colombia. *Nature, Ecology and Evolution*, 1, 3.

Barham, V., Boadway, R., Marchand, M. & Pestieau, P. 1995. Education and the poverty trap. *European Economic Review*, 39, 1257–1275.

Barman, M., Mahapatra, S., Palit, D. & Chaudhury, M. K. 2017. Performance and impact evaluation of solar home lighting systems on the rural livelihood in Assam, India. *Energy for Sustainable Development*, 38, 10–20.

BasicNeeds 2018. Transforming lives of people experiencing mental health problems in the world's poorest communities [Online]. Available: www.basicneeds.org/ [Accessed 19 April 2018].

Bateman, P. W. & Fleming, P. A. 2017. Are negative effects of tourist activities on wildlife over-reported? A review of assessment methods and empirical results. *Biological Conservation*, 211, 10–19.

Battilana, J. & Dorado, S. 2010. Building sustainable hybrid organizations: The case of commercial microfinance organizations. *The Academy of Management Journal*, 53, 1419–1440.

Bebbington, A., Dharmawan, L., Fahmi, E. & Guggenheim, S. 2006. Local capacity, village governance, and the political economy of rural development in Indonesia. *World Development*, 34, 1958–1976.

Beer, A. 2014. Leadership and the governance of rural communities. *Journal of Rural Studies*, 34, 254–262.

Behrman, J. R., Schott, W., Mani, S., Crookston, B. T., Dearden, K., Duc, L. T., Fernald, L. C. H. & Stein, A. D. 2017. Intergenerational transmission of poverty and inequality: Parental resources and schooling attainment and children's human capital in Ethiopia, India, Peru, and Vietnam. *Economic Development and Social Change*, 64(4), 657–697.

Belic, R. (Director) 2011. *Happy*. USA.

Bell, M. L. & Davis, D. L. 2004. A retrospective assessment of mortality from the London smog episode of 1952: The role of influenza and pollution. *Environmental Health Perspectives*, 112, 6–8.

Bemelmans, M., Baert, S., Goemaere, E., Wilkinson, L., Vandendyck, M., Cutsem, G. V., Silva, C., Perry, S., Szumilin, E., Gerstenhaber, R., Kalenga, L., Biot, M. & Ford, N. 2014. Community-supported models of care for people on HIV treatment in sub-Saharan Africa. *Tropical Medicine and International Health*, 19, 968–977.

Benelli, G. & Beier, J. C. 2017. Current vector control challenges in the fight against malaria. *Acta Tropica*, 174, 91–96.

Berg, W., Brunsch, R. & Pazsiczki, I. 2006. Greenhouse gas emissions from covered slurry compared with uncovered during storage. *Agriculture Ecosystems and Environment*, 112, 129–134.

Beschta, R. L. & Ripple, W. J. 2012. The role of large predators in maintaining riparian plant communities and river morphology. *Geomorphology*, 157–158, 88–98.

Beschta, R. L. & Ripple, W. J. 2016. Riparian vegetation recovery in Yellowstone: The first two decades after wolf reintroduction. *Biological Conservation*, 198, 93–103.

Bezboruah, K. C. & Pillai, V. 2015. Exploring the participation of women in financial cooperatives and credit unions in developing countries. *Voluntas*, 26, 913–940.

Bhatti, Y., Taylor, A., Harris, M., Wadge, H., Escobar, E., Prime, M., Patel, H., Carter, A. W., Parston, G., Darzi, A. W. & Udayakumar, K. 2017. Global lessons in frugal innovation to improve health care delivery in the United States. *Global Health Policy*, 36, 1912–1919.

Bickford, N., Smith, L., Bickford, S., Bice, M. R. & Ranglack, D. H. 2017. Evaluating the role of CSR and SLO in ecotourism: Collaboration for economic and environmental sustainability of Arctic resources. *Resources*, 6.

Bird, S. & Hotaling, C. 2017. Multi-stakeholder microgrids for resilience and sustainability. *Environmental Hazards*, 16, 116–132.

Blair, H. 2000. Participation and accountability at the periphery: Democratic local governance in six countries. *World Development*, 28, 21–39.

Blatter, J. 2004. "From spaces of place" to "Spaces of flows"? Territorial and functional governance in cross-border regions in Europe and North America. *International Journal of Urban and Regional Research*, 28, 530–548.

Bluwstein, J. 2017. Creating ecotourism territories: Environmentalities in Tanzania's community-based conservation. *Geoforum*, 83, 101–113.

Bodach, S., Lang, W. & Hamhaber, J. 2014. Climate responsive building design strategies of vernacular architecture in Nepal. *Energy and Buildings*, 81, 227–242.

Bond, T. & Templeton, M. R. 2011. History and future of domestic biogas plants in the developing world. *Energy for Sustainable Development*, 15, 347–354.

Boscolo, D., Tokumoto, P. M., Ferreira, P. A., Ribeiro, J. W. & Santos, J. S. D. 2017. Positive responses of flower visiting bees to landscape heterogeneity depend on functional connectivity levels. *Perspectives in Ecology and Conservation*, 15, 18–24.

Branche, E. 2017. The multipurpose water uses of hydropower reservoir: The SHARE concept. *Comptes Rendus Physique*, 18, 469–478.

Bueren, E. M. V., Bueren, E. T. L. V. & Zijpp, A. J. V. D. 2014. Understanding wicked problems and organized irresponsibility: Challenges for governing the sustainable intensification of chicken meat production. *Current Opinion in Environmental Sustainability*, 8, 1–14.

Calmette, M.-F. & Kilkenny, M. 2012. Rural roads versus African famines. *The Annals of Regional Science*, 49, 373–396.

Center for Global Development 2002. Rich world, poor world: A guide to global development. Washington DC: Center for Global Development.

Center for Health Market Innovations 2018a. *doctHERS* [Online]. Available: https://healthmarketinnovations.org/program/docthers [Accessed 19 April 2018].

Center for Health Market Innovations 2018b. *Doctor from home India* [Online]. Available: https://healthmarketinnovations.org/program/doctor-home-india [Accessed 19 April 2018].

Center for Health Market Innovations 2018c. *MedicallHome* [Online]. Available: https://healthmarketinnovations.org/program/medicallhome [Accessed 19 April 2018].

Centers for Disease Control and Prevention 2015. *Toilets and latrines* [Online]. Atlanta, Georgia. Available: www.cdc.gov/healthywater/global/sanitation/toilets.html [Accessed 8 March 2018].

Chavan, S. B., Uthappa, A. R., Sridhar, K. B., Keerthika, A., Hand, A. K., Newaj, R., Kumar, N., Kumar, D. & Chaturvedi, O. P. 2016. Trees for life: creating sustainable livelihood in Bundelkhand region of central India. *Current Science*, 111, 994–1002.

Cheng, S., LI, Z., Mang, H.-P., Neupane, K., Wauthelet, M. & Huba, E.-M. 2014. Application of fault tree approach for technical assessment of small-sized biogas systems in Nepal. *Applied Energy*, 113, 1372–1381.

Chib, A., Velthoven, M. H. V. & Car, J. 2015. Health adoption in low-resource environments: A review of the use of mobile healthcare in developing countries. *Journal of Health Communication*, 20, 4–34.

Chirenje, L. I. 2017. Contribution of ecotourism to poverty alleviation in Nyanga, Zimbabwe. *Chinese Journal of Population Resources and Environment*, 15, 87–92

Chiu, R. L. 2004. Socio-cultural sustainability of housing: A conceptual exploration. *Housing, Theory and Society*, 21, 65–76.

Choy, Y. K. 2018. Cost-benefit analysis, values, wellbeing and ethics: An indigenous worldview analysis. *Ecological Economics*, 145, 1–9.

Crenna, E., Sozzo, S. & Sala, S. 2018. Natural biotic resources in LCA: Towards an impact assessment model for sustainable supply chain management. *Journal of Cleaner Production*, 172, 3669–3684.

Critchley, W. & Prima, S. D. (Directors) 2014. *Zaï planting pits*. Amsterdam, Netherlands.

Cromley, E. 2008. Cultural embeddedness in vernacular architecture. *Building Research & Information*, 36, 301–304.

Crook, R. C. 2003. Decentralisation and poverty reduction in Africa: The politics of local-central relations. *Public Administration and Development*, 23, 77–88.

Crothers, T. 2012. *The Queen of Katwe: One girl's triumphant path to becoming a chess champion*. New York: Scribner.

Dara, S. 2016. Solidarite Rurale: Transforming agro-education; promoting small farms as gateway for self-employment [Online]. Available: www.changemakers.com/node306408/entries/solidarit%C3%A9-rurale [Accessed 16 January 2018].

Davies, A. B., Tambling, C. J., Kerley, G. I. H. & Asner, G. P. 2016. Limited spatial response to direct predation risk by African herbivores following predator reintroduction. *Ecology and Evolution*, 6, 5728–5748.

Davis, F. R. 2014. *Banned: A history of pesticides and toxicology*. New Haven: Yale University Press.

Dedeke, A. 2017. Creating sustainable tourism ventures in protected areas: An actor-network theory analysis. *Tourism Management*, 61, 161–172.

Dell'Angelo, J., Rulli, M. C. & D'Odorico, P. 2018. The global water grabbing syndrome. *Ecological Economics*, 143, 276–285.

Deshpande, A., Desrochers, A., Ksoll, C. & Shonchoy, A. S. 2017. The impact of a computer-based adult literacy program on literacy and numeracy: Evidence from India. *World Development*, 96, 451–473.

Dessie, A., Alemayehu, E., Mekonen, S., Legesse, W., Kloos, H. & Ambelu, A. 2014. Solar disinfection: An approach for low-cost household water treatment technology in Southwestern Ethiopia. *Journal of Environmental Health Science and Engineering*, 12(1), 25.

Dili, A., Naseer, M. & Varghese, T. Z. 2010. Passive environment control system of Kerala vernacular residential architecture for a comfortable indoor environment: A qualitative and quantitative analyses. *Energy and Buildings*, 42, 917–927.

Dillette, A. K., Douglas, A. C., Martin, D. S. & O'Neill, M. 2017. Resident perceptions on cross-cultural understanding as an outcome of volunteer tourism programs: the Bahamian Family Island perspective. *Journal of Sustainable Tourism*, 25, 1222–1239.

Distefano, T. & Kelly, S. 2017. Are we in deep water? Water scarcity and its limits to economic growth. *Ecological Economics*, 142, 130–147.

Dodo, Y. A., Ahmad, M. H., Dodo, M., Bashir, F. M. & Shika, S. A. 2014. Lessons from Sukur Vernacular Architecture: A Building Material Perspective. *Advanced Materials Research*, 207–210.

Dressler, W. & Roth, R. 2011. The good, the bad, and the contradictory: Neoliberal conservation governance in Rural Southeast Asia. *World Development*, 39, 851–862.

Druk Padma Karpo Educational Society 2016. Druk Padma Karpo School. Available online www.dwls.org/. [Accessed 2 October 2018].

Duby, S. & Engelmeier, T. 2017. Kenya: The world's microgrid lab. Munich, Germany: TFE Consulting.

Duncombe, C., Rosenbaum, S., Hellmann, N., Holmes, C., Wilkinson, L., BIOT, M., Bygone, H., Hoos, D. & Garnett, G. 2015. Reframing HIV care: putting people at the centre of antiretroviral delivery. *Tropical Medicine and International Health*, 20, 430-447.

Dutta, S. 2017. *This is Asia's "Cleanest Village" and it is in India*. Banega Swachh India. Available: https://swachhindia.ndtv.com/asias-cleanest-village-india-5039/ [Accessed 23 July 2018].

Dwivedi, S. L., Bueren, E. T. L. V., Ceccarelli, S., Grando, S., Upadhyaya, H. D. & Ortiz, R. 2017. Diversifying food systems in the pursuit of sustainable food production and healthy diets. *Trends in Plant Science*, 22, 842–856.

Ecobichon, D. J. 2001. Pesticide use in developing countries. *Toxicology*, 160, 27–33.

Embury-Dennis, T. 2017. Costa Rica's electricity generated by renewable energy for 300 days in 2017. *Independent*, 22 November.

Esler, J. 2016. "Green Tibetans" in China: Tibetan geopiety and environmental protection in a multilayered Tibetan landscape. *Asian Ethnicity*, 18, 330–350.

Essuman, A. & Bosumtwi-Sam, C. 2013. School feeding and educational access in rural Ghana: Is poor targeting and delivery limiting impact? *International Journal of Educational Development*, 33, 253–262.

European Network for Rural Development 2016. LEADER local development strategies (LDS): Guidance on design and implementation. Brussels: European Commission.

European Network for Rural Development 2018. *LEADER/CLLD* [Online]. Available: https://enrd.ec.europa.eu/leader-clld_en [Accessed 19 June 2018].

European Parliament and Council 2007. Council regulation 834/2007/EC on organic production and labelling of organic products. *Official Journal*, L189, 1–23.

European Parliament and Council 2009. Directive 2009/128/EC establishing a framework for community action to achieve the sustainable use of pesticides. *Official Journal*, L309, 71–86.

Faiz, A., Faiz, A., Wang, W. & Bennett, C. 2012. Sustainable rural roads for livelihoods and livability. *Procedia – Social and Behavioral Sciences*, 53, 1–8.

Fałkowski, J. 2013. Political accountability and governance in rural areas: Some evidence from the Pilot Programme LEADER+ in Poland. *Journal of Rural Studies*, 32, 70–79.

FAO 2002. *Crops and drops* [Online]. Rome: FAO. Available: www.fao.org/docrep/005/Y3918E/y3918e10.htm [Accessed 15 March 2018].

FAO 2014. FAO's BEFS: Bioenergy and food security approach. FAO.

FAO 2017. *FAOSTAT* [Online]. Rome, Italy: Food and Agriculture Organisation of the United Nations. Available: www.fao.org/faostat/en/#home [Accessed 19 July 2018].

Fischer, B. & Pigneri, A. 2011. Potential for electrification from biomass gasification in Vanuatu. *Energy*, 36, 1640–1651.

Forero-Pineda, C., Escobar-Rodrìguez, D. & Molina, D. 2006. Escuela Nueva's impact on the peaceful social interaction of children in Colombia. *In:* Little, A. W. (ed.), *Education for all and multigrade teaching*. Dordrecht, The Netherlands: Springer.

Forman, R. T. T. 1995. *Land mosaics: The ecology of landscapes and regions*. Cambridge: Cambridge University Press.

Foruzanmehr, A. & Vellinga, M. 2011. Vernacular architecture: questions of comfort and practicability. *Building Research & Information*, 39, 274–285.

France 24 2017. *Dried-up Aral Sea springs back to life* [Online]. YouTube. Available: www.youtube.com/watch?v=pk4v0uu5rkY [Accessed 10 September 2018].

Frazier, M. 1997. *A short history of pest management* [Online]. PennState Extension: PennState. Available: https://extension.psu.edu/a-short-history-of-pest-management [Accessed 15 February 2018].

Fuentes, J. M. 2010. Methodological bases for documenting and reusing vernacular farm architecture. *Journal of Cultural Heritage*, 11, 119–129.

Fukubayashi, Y. & Kimura, M. 2014. Improvement of rural access roads in developing countries with initiative for self-reliance of communities. *Soils and Foundations*, 54, 23–35.

Gardiner, D. T. & Miller, R. W. 2008. *Soils in our environment*. New Jersey: Pearson Prentice Hall.

Gasparatos, A., Doll, C. N. H., Esteban, M., Ahmed, A. & Olang, T. A. 2017. Renewable energy and biodiversity: Implications for transitioning to a Green Economy. *Renewable and Sustainable Energy Reviews*, 70, 161–184.

Gautam, A. 2008. *Climate responsive vernacular architecture: Jharkhand, India*. Manhattan, KS: Kansas State University.

Gautam, S., Schreinemachers, P., Uddin, M. N. & Srinivasan, R. 2017. Impact of training vegetable farmers in Bangladesh in integrated pest management (IPM). *Crop Protection*, 102, 161–169.

Ghaffarian Hoseini, A. H., Dahlan, N. D., Berardi, U., Ghaffarian Hoseini, A. & Makaremi, N. 2013. The essence of future smart houses: From embedding ICT to adapting to sustainability principles. *Renewable and Sustainable Energy Reviews*, 24, 593–607.

Gippner, O., Dhakal, S. & Sovacool, B. K. 2012. Microhydro electrification and climate change adaptation in Nepal: Socioeconomic lessons from the Rural Energy Development Program (REDP). *Mitigation and Adaptation Strategies for Global Change*, 18, 407–427.

Gkartzios, M. & Scott, M. 2010. Residential mobilities and house building in rural Ireland: Evidence from three case studies. *Sociologia Ruralis*, 50, 64–84.

Glewwe, P. & Jacoby, H. G. 2004. Economic growth and the demand for education: Is there a wealth effect? *Journal of Development Economics*, 74, 33–51.

Good, A. G. & Beatty, P. H. 2011. Fertilizing nature: A tragedy of excess in the commons. *PLoS Biology*, 9, 1–9.

Green, R. 2005. Community perceptions of environmental and social change and tourism development on the island of Koh Samui, Thailand. *Journal of Environmental Psychology*, 25, 37–56.

Grgić, I., Žimbrek, T., Tratnik, M., Markovina, J. & Juračak, J. 2010. Quality of life in rural areas of Croatia: To stay or to leave? *African Journal of Agricultural Research*, 5, 653–660.

Grimes, C. E., Billingsley, M. L., Dare, A. J., Day, N., George, P. M., Kamara, T. B., Mkanawire, N. C., Leather, A. & Lavy, C. B. D. 2015. The demographics of patients affected by surgical disease in district hospitals in two sub-Saharan African countries: A retrospective descriptive analysis. *Springer Plus*, 4, 9.

Grydehøj, A. & Kelman, I. 2017. The eco-island trap: Climate change mitigation and conspicuous sustainability. *Area*, 49, 106–113.

Gunnerson, C. C. & Stuckey, D. C. 1986. Anaerobic digestion: Principles and practices for biogas systems. *Integrated Resource Recovery*. Washington DC: The World Bank.

Guo, T. & Englehardt, J. D. 2015. Principles for scaling of distributed direct potable water reuse systems: A modeling study. *Water Research*, 75, 146–163.

Haddad, N. M., Brudvig, L. A., Clobert, J., Davies, K. F., Gonzalez, A., Holt, R. D., Lovejoy, T. E., Sexton, J. O., Austin, M. P., Collins, C. D., Cook, W. M., Damschen, E. I., Ewers, R. M., Foster, B. L., Jenkins, C. N., King, A. J., Laurance, W. F., Levey, D. J., Margules, C. R., Melbourne, B. A., Nicholls, A. O., Orrock, J. L., Song, D.-X. & Townshend, J. R. 2015. Habitat fragmentation and its lasting impact on Earth's ecosystems. *Science Advances*, 1, 9.

Hajjar, R., Badini, O. S. & Kozak, R. A. 2017. Promoting small and medium forest enterprises in national REDD+ strategies: A global analysis of enabling environments. *Climate Policy*, 17, 731–763.

Han, Z., Dan, Z., Shi, G., Shen, L., Xu, W. & Xie, Y. 2015. Characteristics and management of domestic waste in a rural area of the Tibetan Plateau. *Journal of the Air and Waste Management*, 65, 1365–1375.

Hansjürgens, B., Schröter-Schlaack, C., Berghöfer, A. & Lienhoop, N. 2017. Justifying social values of nature: Economic reasoning beyond self-interested preferences. *Ecosystem Services*, 23, 9–17.

Hargreaves, J. R., Boccia, D., Evans, C. A., Adata, M., Petticrew, M. & Porter, J. D. H. 2011. The social determinants of tuberculosis: From evidence to action. *American Journal of Public Health*, 101, 654–662.

Harland, M. 2017. Permaculture: Tools for making women's lives more abundant. *Feminist Theology*, 25, 240–247.

Henseler, M. & Maisonnave, H. 2018. Low world oil prices: A chance to reform fuel subsidies and promote public transport? A case study for South Africa. *Transportation Research Part A*, 108, 45–62.

Hermes, N., Lensink, R. & Meesters, A. 2011. Outreach and efficiency of microfinance institutions. *World Development*, 39, 938–948.

Hernández-Aguilar, J. A., Cortina-Villar, H. S., García-Barrios, L. E. & Castillo-Santiago, M. Á. 2017. Factors limiting formation of community forestry enterprises in the southern Mixteca region of Oaxaca, Mexico. *Environmental Management*, 59, 490–504.

Hertel, T. W. & Baldos, U. L. C. 2016. Attaining food and environmental security in an era of globalization. *Global Environmental Change*, 41, 195–205.

Hoffman, S. L., Vekemans, J., Richie, T. L. & Duffy, P. E. 2015. The march towards malaria vaccines. *American Journal of Preventative Medicine*, 49, S319–S333.

Holmes, J., Jones, B. & Heap, B. 2015. Smart villages. *Science*, 350(6259), 359.

Hoornweg, D. & Bhada-Tata, P. 2012. What a waste: A global review of solid waste management. *In:* Bank, T. W. (ed.), *Urban Development Knowledge Papers*. Washington DC: The World Bank.

Hudelson, C. & Cluver, L. 2015. Factors associated with adherence to antiretroviral therapy among adolescents living with HIV/AIDS in low- and middle-income countries: a systematic review. *AIDS Care*, 27, 805–816.

Hunger Notes 2016. *2016 world hunger and poverty facts and statistics* [Online]. Washington DC: Hunger Notes: World Hunger Education Service. Available: www.worldhunger.org/2015-world-hunger-and-poverty-facts-and-statistics/ [Accessed 15 February 2018].

Idris, N. H., Osman, M. J., Kanniah, K. D., Idris, N. H. & Ishak, M. H. I. 2016. Engaging indigenous people as geo-crowdsourcing sensors for ecotourism mapping via mobile data collection: A case study of the Royal Belum State Park. *Cartography and Geographic Information Science*, 44, 113–127.

Ihekire, O., Salawu, S. A. I. & Opadele, T. 2010. Causes of hand injuries in a developing country. *Canadian Journal of Surgery*, 53, 161–166.

Indraganti, M. 2010. Understanding the climate sensitive architecture of Marikal, a village in Telangana region in Andhra Pradesh, India. *Building and Environment*, 45, 2709–2722.

Islar, M., Brogaard, S. & Lemberg-Pedersen, M. 2017. Feasibility of energy justice: Exploring national and local efforts for energy development in Nepal. *Energy Policy*, 105, 668–676.

Jack, C. 2014. Integrated Pest Management. University of Liverpool.

Jagnoor, J., Prinja, S., Lakshmi, P. V. M., Aggarwal, S., Gabbe, B. & Ivers, R. Q. 2018. The impact of road traffic injury in North India: A mixed-methods study protocol. *BMJ Open*, 5, 5.

Johnston, D. C. 2007. These roads were made for walking? The nature and use of rural public transport services in Garut Regency, West Java, Indonesia. *Singapore Journal of Tropical Geography*, 28, 171–187.

Jolin, L. 2018. Adventures in life sciences. *Imperial*, 43, 46–47. Imperial College, London: Imperial College, London.

Joon, V., Chandra, A. & Bhattacharya, M. 2009. Household energy consumption pattern and socio-cultural dimensions associated with it: A case study of rural Haryana, India. *Biomass and Bioenergy*, 33, 1509–1512.

Kanase-Patil, A. B., Saini, R. P. & Sharma, M. P. 2010. Integrated renewable energy systems for off grid rural electrification of remote area. *Renewable Energy*, 35, 1342–1349.

Kaza, N. 2010. Understanding the spectrum of residential energy consumption: A quantile regression. *Energy Policy*, 38, 6574–6585.

Kazeem, A. & Musalia, J. M. 2017. The implications of ethnicity, gender, urban-rural residence, and socioeconomic status for progress through school among children in Nigeria. *Social Indicators Research*, 132, 861–884.

Khoury, C. K., Bjorkman, A. D., Dempewolf, H., Ramirez-Villegas, J., Guarino, L., Jarvis, A., Rieseberg, L. H. & Struik, P. C. 2014. Increasing homogeneity in global food supplies and the implications for food security. *Proceedings of the National Academy of Sciences*, 111, 4001–4006.

Kilkenny, M. 2010. Urban/regional economics and rural development. *Journal of Regional Science*, 50, 449–470.

Kimura, K.-I. 1994. Vernacular technologies applied to modern architecture. *Renewable Energy*, 5, 900–907.

Kirk, J. 2001. Media and social change: The modernizing influences of television in rural India. *Media, Culture & Society*, 23, 147–169.

Kumar, A., Tewari, A., Shroff, G., Chittamuru, D., Kam, M. & Canny, J. 2010. An exploratory study of unsupervised mobile learning in rural India. *Proceedings of the SIGCHI Conference on Human Factors in Computing Systems*. Atlanta, Georgia: ACM.

Lai, C. S., Jia, Y., Xu, Z., Lai, L. L., Li, X., Cao, J. & McCulloch, M. D. 2017. Levelized cost of electricity for photovoltaic/biogas power plant hybrid system with electrical energy storage degradation costs. *Energy Conversion and Management*, 153, 34–47.

Laidin, A. Z. 2015. Smart villages – The Malaysian approach. *In:* Holmes, J. et al. (eds.), *Smart Villages: New Thinking for Off-Grid Communities Worldwide*, 42.

Laitinen, J., Moliis, K. & Surakka, M. 2017. Resource efficient wastewater treatment in a developing area – Climate change impacts and economic feasibility. *Ecological Engineering*, 103, 217–225.

Lam, N. L., Smith, K. R., Gauthier, A. & Bates, M. N. 2012. Kerosene: A review of household uses and their hazards in low and middle income countries. *Journal of Toxicology and Environmental Health Part B, Critical Reviews*, 15, 396–432.

Lamb, W. F. 2016. Which countries avoid carbon-intensive development? *Journal of Cleaner Production*, 131, 523–533.

Landers, T. F., Cohen, B., Wittum, T. E. & Larson, E. L. 2012. A review of antibiotic use in food animals: Perspective, policy, and potential *Public Health Reports*, 127, 4–22.

Laxmi, V., Parikh, J., Karmakar, S. & Dabrase, P. 2003. Household energy, women's hardship and health impacts in rural Rajasthan, India: Need for sustainable energy solutions. *Energy for Sustainable Development*, 7, 50–68.

Lee, C.-H. & Wang, C.-H. 2017. Estimating residents' preferences of the land use program surrounding Forest Park, Taiwan. *Sustainability*, 9.

Lee, T. H. & Hsieh, H.-P. 2016. Indicators of sustainable tourism: A case study from a Taiwan's wetland. *Ecological Indicators*, 67, 779–787.

Legros, G., Rijal, K. & Seyedi, B. 2011. Decentralized energy access and the millennium development goals: An analysis of the development benefits of micro-hydropower in rural Nepal. Rugby, UK: United Nations Development Program.

Lim, R. M. 2007. Cultural sustainability and development: Drukpa and Burman vernacular architecture. *In*: Zetter, R. & Butina Watson, G. (eds.), *Designing Sustainable Cities in the Developing World*. Farnham: Ashgate, 21.

Little, A. 2006. *Education for all and multigrade teaching: Challenges and opportunities*. Dordrecht, The Netherlands: Springer.

Liu, X. & Hannum, E. 2017. Early poverty exposure predicts young adult educational outcomes in China. *China Economic Review*, 44, 79–97.

Lo, S., Das, P. & Horton, R. 2016. Early childhood development: The foundation of sustainable development. *Lancet*, 389, 9–11.

López-González, A., Domenech, B., Gómez-Hernández, D. & Ferrer-Martí, L. 2017. Renewable microgrid projects for autonomous small-scale electrification in Andean countries. *Renewable and Sustainable Energy Reviews*, 79, 1255–1265.

Lopez, S. L. 2010. The remittance house: Architecture of migration in rural Mexico. *Buildings & Landscapes: Journal of the Vernacular Architecture Forum*. Minneapolis: University of Minnesota Press, 33–52.

Lorimer, J., Sandom, C., Jepson, P., Doughty, C., Barua, M. & Kirby, K. J. 2015. Rewilding: Science, practice and politics. *Annual Review of Environment and Resources*, 40, 39–62.

Luby, J., Belden, A., Botteron, K., Marrus, N., Harms, M. P., Babb, C., Nishino, T. & Barch, D. 2013. The effects of poverty on childhood brain development: The mediating effect of caregiving and stressful life events. *JAMA Pediatrics*, 167, 1135–1142.

Lucas, K. 2011. Making the connections between transport disadvantage and the social exclusion of low income populations in the Tshwane Region of South Africa. *Journal of Transport Geography*, 19, 1320–1334.

Lucon, O., Ürge-Vorsatz, D., Ahmed, A. Z., Akbari, H., Bertoldi, P., Cabeza, L. F., Eyre, N., GadgilL, A., Harvey, L. D. D., Jiang, Y., Liphoto, E., Mirasgedis, S., Murakami, S., Parikh, J., Pyke, C. & Vilariño, M. V. 2014. Buildings. *In:* Edenhofer, O., Pichs-Madruga, R., Sokona, Y., Farahani, E., Kadner, S., Seyboth, K., Adler, A., Baum, I., Brunner, S., Eickemeier, P., Kriemann, B., Savolainen, J., Schlömer, S., Stechow, C. V., Zwickel, T. & Minx, J C. (eds.), *Climate Change 2014: Mitigation of climate change.* Cambridge and New York: Contribution of Working Group III to the Fifth Assessment Report of the Intergovernmental Panel on Climate Change.

Lyngdoh, S., Mathur, V. B. & Sinhaa, B. C. 2017. Tigers, tourists and wildlife: Visitor demographics and experience in three Indian Tiger Reserves. *Biodiversity and Conservation*, 26, 2187–2204.

Lyon, A., Hunter-Jones, P. & Warnaby, G. 2017. Are we any closer to sustainable development? Listening to active stakeholder discourses of tourism development in the Waterberg Biosphere Reserve, South Africa. *Tourism Management*, 61, 234–247.

Macinko, J. & Harris, M. J. 2015. Brazil's family health strategy – Delivering community-based primary care in a universal health system. *The New England Journal of Medicine*, 372, 2177–2181.

Mahapatra, S. & Dasappa, S. 2012. Rural electrification: Optimising the choice between decentralised renewable energy sources and grid extension. *Energy for Sustainable Development*, 16, 146–154.

Makombe, G., Namara, R. E., Awulachew, S. B., Hagos, F., Ayana, M. & Kanjere, M. 2017. An analysis of the productivity of technical efficiency of smallholder irrigation in Ethiopia. *Water SA*, 43, 48–57.

Malvestio, A. C., Fischer, T. B. & Montaño, M. 2018. The consideration of environmental and social issues in transport policy, plan and programme making in Brazil: A systems analysis. *Journal of Cleaner Production*, 179, 674–689.

Maphatsoe, K. 2017. Community is uplifted in Actonville. *Benoni City Times*. 23 October.

Mara, D. D. 1985. Ventilated improved pit latrines: Guidelines for the selection of design options. *In:* Program, U. N. D. (ed.), *Technical Advisory Group Discussion Paper 4*. Washington DC: The World Bank.

Marsh, T. L., Yoder, J., Deboch, T., McElwain, T. F. & Palmer, G. H. 2016. Livestock vaccinations translate into increased human capital and school attendance by girls. *Science Advances*, 2.

Masters, B. (Director) (2017). *50 years ago, this was a wasteland. He changed everything*. YouTube: National Geographic. Available: www.youtube.com/watch?v=ZSPkcpGmflE [Accessed 13 September 2018].

Mathur, V. B., Verma, A., Dudley, N., Stolton, S., Hockings, M. & James, R. 2006. Opportunities and challenges for Kaziranga National Park, Assam over the next fifty years. UNESCO.

Maudlin, D. 2010. Crossing boundaries: Revisiting the thresholds of vernacular architecture. *Vernacular Architecture*, 41, 10–14.

Maxwell, A. 2016. Post-HydroAysen, Chile's rivers remain at risk. Available from: www.nrdc.org/experts/amanda-maxwell/post-hidroaysen-chiles-rivers-remain-risk [Accessed 20 February 2018].

Mayett-Moreno, Y., Villarraga-Floréz, L. F. & Rodríguez-Piñeros, S. 2017. Young farmers' perceptions about forest management for ecotourism as an alternative for development in Puebla, Mexico. *Sustainability*, 9.

McAndrews, C., Beyer, K., Guse, C. & Layde, P. 2017. Linking transportation and population health to reduce racial and ethnic disparities in transportation injury: Implications for practice and policy. *International Journal of Sustainable Transportation*, 11, 197–205.

McMaster, W. D. (Director) 2014. *Forest man*. YouTube [Online]. Available: www.youtube.com/watch?v=HkZDSqyE1do. [Accessed 15 September 2018].

MDG Monitor. 2017. *MDG 2: Achieve universal primary education* [Online]. United Nations. Available: www.mdgmonitor.org/mdg-2-achieve-universal-primary-education/ [Accessed 19 July 2018].

Mie, A., Andersen, H. R., Gunnarsson, S., Kahl, J., Kesse-Guyot, E., Rembiałkowska, E., Quaglio, G. & Grandjean, P. 2017. Human health implications of organic food and organic agriculture: A comprehensive review. *Environmental Health*, 16, 22.

Mihai, F.-C. & Ingrao, C. 2018. Assessment of biowaste losses through unsound waste management practices in rural areas and the role of home composting. *Journal of Cleaner Production*, 172, 1631–1638.

Mohee, R. 2007. Waste management opportunities for rural communities. *In:* FAO (ed.), *Agricultural and Food Engineering Working Document*. Rome: Food and Agricultural Organization of the United Nations.

Mokyr, J. 2017. Great famine. *Encyclopaedia Britannica*.

Munang, R. & Mgendi, R. 2015. Africa's broken food security value chain: unravelling the hidden fortune. *This is Africa*. 20 August. Available: https://thisisafrica.me/africas-broken-food-security-value-chain-unravelling-hidden-fortune/ [Accessed 11 September 2018].

Murdock, L. L. & Baoua, I. B. 2014. On Purdue improved cowpea storage (PICS) technology: Background, mode of action, future prospects. *Journal of Stored Products Research*, 58, 3–11.

Murtaza, G., Usman, Y., Niazi, N. K., Usman, M. & Hussain, T. 2017. Bioaccumulation of potentially toxic elements in cereal and legume crops: A review. *CLEAN – Soil, Air, Water*, 45, 15.

Mvukiyehe, E. & Samii, C. 2017. Promoting democracy in fragile states: Field experimental evidence from Liberia. *World Development*, 95, 254–267.

Mwase, N. 1996. Developing an environment-friendly transport system in Tanzania: Some policy considerations. *Transport Reviews*, 16, 145–156.

Mwase, N. 2003. The liberalisation, de-regulation and privatisation of the transport sector in sub-Saharan Africa: Experiences, challenges and opportunities. *Journal of African Economies*, 12, ii153–ii192.

Nasir, M., Khan, H. A., Hussain, A., Mateen, L. & Zaffar, N. A. 2018. Solar PV-based scalable DC microgrid for rural electrification in developing regions. *IEEE Transactions on Sustainable Energy*, 9, 390–399.

Nathan, S. B. S. & Rosli, M. M. 2014. Distributional effects of non-farm incomes in a Malaysian rice bowl. *International Journal of Social Economics*, 43, 205–220.

Nepal, S. K. 2007. Tourism and rural settlements Nepal's Annapurna region. *Annals of Tourism Research*, 34, 855–875.

New Economics Foundation 2016. *Happy Planet Index 2016 Methods paper* [Online]. New Economics Foundation. Available: https://static1.squarespace.com/static/5735c421e321402778ee0ce9/t/578dec7837c58157b929b3d6/1468918904805/Methods+paper_2016.pdf [Accessed 29 March 2018].

New Economics Foundation 2017. *Costa Rica* [Online]. New Economics Foundation. Available: http://happyplanetindex.org/countries/costa-rica [Accessed 29 March 2018].

Newbold, T., Hudson, L. N., Hill, S. L. L., Contu, S., Lysenko, I., Senior, R. A., Börger, L. et al. 2015. Global effects of land use on local terrestrial biodiversity. *Nature*, 520, 45–50.

Nguyen, A.-T., Tran, Q.-B., Tran, D.-Q. & Reiter, S. 2011. An investigation on climate responsive design strategies of vernacular housing in Vietnam. *Building and Environment*, 46, 2088–2106.

Nogrady, B. 2017. *Tasmanian irrigators pioneer high-tech, localised water management* [Online]. CSIRO. Available: https://blogs.csiro.au/ecos/sense-t/ [Accessed 14 June 2018].

Nonoyama-Tarumi, Y. & Ota, Y. 2010. Early childhood development in developing countries: Pre-primary education, parenting, and health care. *In:* UNESCO (ed.), *Paper commissioned for the EFA Global Monitoring Report 2011, The hidden crisis: Armed conflict and education.* New York: UNESCO.

O'Connor, J. 2017. *Battery showdown: Lead-acid vs. Lithium-Ion* [Online]. Medium. Available: https://medium.com/solar-microgrid/battery-showdown-lead-acid-vs-lithium-ion-1d37a1998287 [Accessed 20 March 2018].

Odell, K. 2010. Measuring the impact of microfinance: Taking another look. *Grameen Foundation Publication Series.* Washington DC: Grameen Foundation.

Ohandja, D.-G., Donovan, S., Castle, P. & Voulvoulis, N. 2012. Regulatory systems and guidelines for the management of risk. *In:* Plant, J. A., Voulvoulis, N. & Ragnarsdottir, K. V. (eds.), *Pollutants, human health and the environment: A risk based approach.* London: Wiley-Blackwell.

Orenstein, D. E. & Shach-Pinsley, D. 2017. A comparative framework for assessing sustainability initiatives at the regional scale. *World Development*, 98, 245–256.

Organisation for Economic Co-operation and Development 2016. OECD regional well-being: A user's guide. Paris: OECD.

Pai, M., Behr, M. A., Dowdy, D., Dheda, K., Divangahi, M., Boehme, C. C., Ginsberg, A., Swaminathan, S., Spigelman, M., Getahun, H., Menzies, D. & Raviglione, M. 2016. Tuberculosis. *Nature Review: Disease Primers*, 2, 1–23.

Painter, L. E., Beschta, R. L., Larsen, E. J. & Ripple, W. J. 2015. Recovering aspen follow changing elk dynamics in Yellowstone: Evidence of a trophic cascade? *Ecology*, 96, 252–263.

Pal, S. & Wahhaj, Z. 2017. Fiscal decentralisation, local institutions and public good provision: Evidence from Indonesia. *Journal of Comparative Economics*, 45, 383–409.

Palit, D. & Chaurey, A. 2011. Off-grid rural electrification experiences from South Asia: Status and best practices. *Energy for Sustainable Development*, 15, 266–276.

Papanikolaou, A. D., Kühn, I., Freznel, M., Kuhlmann, M., Poschlod, P., Potts, S. G., Roberts, S. P. M. & Schweiger, O. 2017. Wild bee and floral diversity co-vary in response to the direct and indirect impacts of land use. *Ecosphere*, 8, 14.

Parliament and Council of the European Union 2008. Directive 2008/98/EC on waste. *Official Journal* L 312, 3–30.

Pauli, J. 2008. A house of one's own: Gender, migration, and residence in rural Mexico. *American Ethnologist*, 35, 171–187.

Peisker, A. & Dalai, S. 2015. Data analytics for rural development. *Indian Journal of Science and Technology*, 8, 50–60.

Pemberton, S. & Goodwin, M. 2010. Rethinking the changing structures of rural local government – State power, rural politics and local political strategies. *Journal of Rural Studies*, 26, 272–283.

Pennise, D. 2005. *Biomass pollution basics*. Berkeley, California: University of California, Berkeley.

Pérez-Escamilla, R. & Moran, V. H. 2017. The role of nutrition in integrated early child development in the 21st century: Contribution from the Maternal and Child Nutrition Journal. *Maternal and Child Nutrition*, 13.

Pergola E. M., Persiani, A., Palese, A. M., Meo, V. D., Pastore, V., D'Adamo, C. & Celano, G. In Press. Composting: The way for a sustainable agriculture. *Applied Soil Ecology*.

Peterson, R. K. D., Varella, A. C. & Higley, L. G. 2017. Tolerance: The forgotten child of plant resistance. *PeerJ*.

Pinard, M. I., Newport, S. J. & Rijn, J. V. 2016. Addressing the road maintenance challenge in Africa: What can we do to solve this continuing problem? *International Conference on Transport and Road Research*. Whitesands Hotel, Mombasa, Kenya.

Poggetti, R. S. 2008. Acute care surgeon South American model. *World Journal of Surgery*, 32, 1626–1629.

Pretty, J., Toulmin, C. & Williams, S. 2011. Sustainable intensification in African agriculture. *International Journal of Agricultural Sustainability*, 9, 5–24.

Primdahl, J., Kristensen, L. S. & Swaffield, S. 2013. Guiding rural landscape change: Current policy approaches and potentials of landscape strategy making as a policy integrating approach. *Applied Geography*, 42, 86–94.

Prinsloo, G., Mammoli, A. & Dobson, R. 2017. Customer domain supply and load coordination: A case for smart villages and transactive control in rural off-grid microgrids. *Energy*, 135, 430–441.

Ragunathan, M. V. & Sultana, M. F. 2017. Smart Cities & Smart Villages: A mission to transform the nation. *In:* Ragunathan, M. V. & Sultana, M. F. (eds.), National

Conference on Smart Cities and Smart Villages: A mission to transform the nation, Kodaikanal. *Shanlax International Journal of Management*.

Randell, H. & Gray, C. 2016. Climate variability and educational attainment: Evidence from rural Ethiopia. *Global Environmental Change*, 41, 11–123.

Ranjan, R. 2017. Preserving ecologically sensitive hotspots under endogenous regulation and valuation constraints. *Natural Resource Modeling*, 30, 23.

Ranson, H. & Lissenden, N. 2016. Insecticide resistance in African *Anopheles* mosquitoes: A worsening situation that needs urgent action to maintain malaria control. *Trends in Parasitology*, 32, 187–196.

Rapoport, A. 1987. On the cultural responsiveness of architecture. *Journal of Architectural Education*, 41, 10–15.

Rapoport, A. 1998. Using "Culture" in housing design. *Housing and Society*, 25, 1–20.

Ray, S., Ghosh, B., Bardhan, S. & Bhattacharyya, B. 2016. Studies on the impact of energy quality on human development index. *Renewable Energy*, 92, 117–126.

Richter, L. M., Daelmans, B., Lombardi, J., Heymann, J., Boo, F. L., Behrman, J. R., Lu, C., Lucas, J. E., Pérez-Escamilla, R., Dua, T., Bhutta, Z. A., Stenberg, K., Gertler, P. & Darmstadt, G. L. 2017. Investing in the foundation of sustainable development: Pathways to scale up for early childhood development. *Lancet*, 389, 103–118.

Riginos, C. 2015. Climate and the landscape of fear in African savanna. *Journal of Animal Ecology*, 84, 124–133.

Rönkkö, E., Luusa, A., Aarrevaara, E., Herneoja, A. & Muilu, T. 2017. New resource-wise planning strategies for smart urban-rural development in Finland. *Systems*, 5, 12.

Roopnarain, A. & Adeleke, R. 2017. Current status, hurdles and future prospects of biogas digestion technology in Africa. *Renewable and Sustainable Energy Reviews*, 67, 1162–1179.

Roth, G. W. 1996. Crop rotations and conservation tillage. University Park, Pennsylvania: The Pennsylvania State University.

Rowse, L. E. 2011. Design of small scale anaerobic digesters for application in rural developing countries. Master of Science, University of South Florida.

Rowsell, J., Morrell, E. & Alvermann, D. E. 2017. Confronting the digital divide: Debunking brave new world discourses. *The Reading Teacher*, 71, 157–165.

Royal Flying Doctor Service (2017) About the RFDS. Available: www.flyingdoctor.org.au/about-the-rfds/ [Accessed 7 December 2017].

Royte, E. 2017. Nearly a billion people still defecate outdoors. Here's why. *National Geographic*.

Rubagiza, J., Were, E. & Sutherland, R. 2011. Introducing ICT into schools in Rwanda: Educational challenges and opportunities. *International Journal of Educational Development*, 31, 37–43.

Rurak, M. 2012. *Advantages and disadvantages of controlled-release fertilisers*. Madison, Wisconsin: Department of Soil Science, University of Wisconsin, Madison.

Ryser, L. & Halseth, G. 2012. Resolving mobility constraints impeding rural seniors' access to regionalized services. *Journal of Aging and Social Policy*, 24, 328–344.

Sai, S. S. T. 2016. Smart Villages – Need of emerging India. *International Journal Of Innovative Research in Information Security*, 9.

Santos, C. F. D., Acosta, A. L., Dorneles, A. L., Santos, P. D. S. D. & Blochtein, B. 2016. Queens become workers: Pesticides alter caste differentiation in bees. *Scientific Reports* 6, 9.

Scharf, R. J., Maphula, A., Pullen, P. C., Shrestha, R., Matherne, G. P., Roshan, R. & Koshy, B. 2017. Global Disability: Empowering children of all abilities. *Pediatrics Clinics of North America*, 64, 769–784.

Schreinemachers, P., Chen, H.-P., Nguyen, T. T. L., Buntong, B., Bouapao, L., Gautam, S., Le, N. T., Pinn, T., Vilaysone, P. & Srininvasan, R. 2017. Too much to handle? Pesticide dependence of smallholder vegetable farmers in Southeast Asia. *Science of the Total Environment*, 593–594, 470–477.

Schreinemachers, P. & Tipraqsa, P. 2012. Agricultural pesticides and land use intensification in high, middle and low income countries. *Food Policy*, 37, 616–626.

Secretariat for the Convention on Biological Diversity 2000. Sustaining life on earth: How the convention on biological diversity promotes nature and human well-being. *The Interim Secretariat for the Convention on Biological Diversity.* Montreal, Canada: Secretariat of the Convention on Biological Diversity, United Nations Environment Programme.

Seyfrit, C. L., Bjarnason, T. & Olafsson, K. 2010. Migration intentions of rural youth in Iceland: Can a large-scale development project stem the tide of outmigration? *Society and Natural Resources*, 23, 1201–1215.

Shafiee, S. & Topal, E. 2009. When will fossil fuel reserves be diminished? *Energy Policy*, 37, 181–189.

Shastry, V., Mani, M. & Tenorio, R. 2014. Impacts of modern transitions on thermal comfort in vernacular dwellings in warm-humid climate of Sugganahalli (India). *Indoor and Built Environment*, 23, 543–564.

Shrestha, J. K., Benta, A., Lopes, R. B. & Lopes, N. 2014. A multi-objective analysis of a rural road network problem in the hilly regions of Nepal. *Transportation Research Part A*, 64, 43–53.

Siddaiah, R. & Saini, R. P. 2016. A review on planning, configurations, modeling and optimization techniques of hybrid renewable energy systems for off grid applications. *Renewable and Sustainable Energy Reviews*, 58, 376–396.

Siemons, R. V. 2001. Identifying a role for biomass gasification in rural electrification in developing countries: The economic perspective. *Biomass and Bioenergy*, 20, 271–285.

Silva, A. P. M. D., Schweizer, D., Marques, H. R., Teixeira, A. M. C., Santos, T. V. M. N. D., Sambuichi, R. H. R., Badari, C. G., Gaudare, U. & Brancalion, P. H. S. 2017. Can current native tree seedling production and infrastructure meet an increasing forest restoration demand in Brazil? *Restoration Ecology*, 25, 509–515.

Singh, J., Panesar, B. S. & Sharma, S. K. 2008. Energy potential through agricultural biomass using geographical information system – A case study of Punjab. *Biomass and Bioenergy*, 32, 301–307.

Singh, M. K., Mahapatra, S. & Atreya, S. 2009. Bioclimatism and vernacular architecture of north-east India. *Building and Environment*, 44, 878–888.

Singh, M. K., Mahapatra, S. & Atreya, S. 2011. Solar passive features in vernacular architecture of North-East India. *Solar Energy*, 85.

Singh, S., Singh, M. & Kaushik, S. C. 2016. Feasibility study of an islanded microgrid in rural area consisting of PV, wind, biomass and battery energy storage system. *Energy Conversion and Management*, 128, 178–190.

Smith, S. R. & Jasim, S. 2009. Small-scale home composting of biodegradable household waste: Overview of key results from a 3-year research programme in West London. *Waste Management and Research*, 27, 941–950.

Soto, B., Munthali, S. M. & Breen, C. 2001. Perceptions of the Forestry and Wildlife Policy by the local communities living in the Maputo Elephant Reserve, Mozambique. *Biodiversity and Conservation*, 10, 1723–1738.

Source Trace 2018. *How mobile apps are helping agriculture in achieving sustainable development* [Online]. Available: www.sourcetrace.com/mobile-apps-for-agriculture/ [Accessed 19 June 2018].

Sovacool, B. K. & Dworkin, M. H. 2015. Energy justice: Conceptual insights and practical applications. *Applied Energy*, 142, 435–444.

Stamoulis, K. & Lipper, L. 2012. Agricultural input subsidies and the green economy: Fertilizer subsidies in Sub-Saharan Africa. Rome, Italy: Food and Agriculture Organisation of the United Nations.

Stauffer, B. 2011. *Overhung latrine* [Online]. Willisau, Switzerland: Sustainable Sanitation and Water Management Toolbox. Available: www.sswm.info/water-nutrient-cycle/waste-water-treatment/hardwares/user-interface/overhung-latrine [Accessed 27 March 2018]

Steyn, W. J. V., Nokes, B., Plessis, L. D., Agacer, R., Burmas, N. & Popescu, L. 2015. Evaluation of the effect of rural road condition on agricultural produce transportation. *Transportation Research Record*, 2473, 33–41.

Struick, P. C., Kuyper, T. W., Brussaard, L. & Leeuwis, C. 2014. Deconstructing and unpacking scientific controversies in intensification and sustainability: Why the tensions in concepts and values? *Current Opinion in Environmental Sustainability*, 8, 80–88.

SunSmart 2008. Skin cancer prevention: A blue chip investment in Victoria. Victoria: SunSmart Cancer Council.

Supradist, N. 2004. *Economic leakage in tourism sector*. Master of Science in Environmental Management and Policy, Lund University.

Surendra, K. C., Takara, D., Hashimoto, A. G. & Khanal, S. K. 2014. Biogas as a sustainable energy source for developing countries: Opportunities and challenges. *Renewable and Sustainable Energy Reviews*, 31, 846–859.

Tasci, B. G. 2015. "Sustainability" education by sustainable school design. *Procedia – Social and Behavioral Sciences*, 186, 868–873.

Teenstra, E., Vellinga, T., Aektasaeng, N., Amatayakul, W., Ndambi, A., Pelster, D., Germer, L., Jenet, A., Opio, C. & Andeweg, K. 2014. Global assessment of manure management policies and practices. Wageningen: Wageningen UR Livestock Research.

Titley, M. A., Snaddon, J. L. & Turner, E. C. 2017. Scientific research on animal biodiversity is systematically biased towards vertebrates and temperate regions. *PLoS One*, 12, 14.

Tiwari, P. 2001. Housing and development objectives in India. *Habitat International*, 25, 229–253.

Turner, T. (2015) Druk dragon garden in Ladakh: The design and concept in 2015. Presentation to Arup Associates. Available: www.youtube.com/watch?v=3l4Cosg6eqs [Accessed 2 October 2018].

US EPA 2017. *Understanding global warming potentials* [Online]. Washington DC. Available: www.epa.gov/ghgemissions/understanding-global-warming-potentials [Accessed 3 May 2018].

US Geological Survey 2016. *The water in you* [Online]. Washington DC. Available: https://water.usgs.gov/edu/propertyyou.html [Accessed 20 July 2018].

Uduku, O. 2015. Designing schools for quality: An international, case study-based review. *International Journal of Educational Development*, 44, 56–64.

UN-REDD. 2016. *UN-REDD programme* [Online]. United Nations. Available: www.un-redd.org/ [Accessed 27 March 2018].

UNEP 2004. Guidelines on municipal wastewater management. The Hague, The Netherlands.

UNESCO 1990. World declaration on education for all. Jomtien, Thailand.

UNESCO & SCOPE 2007. Human alteration of the nitrogen cycle: Threats, benefits and opportunities. Paris, France: UNESCO-SCOPE.

United Nations 1948. The universal declaration of human rights. Paris.

United Nations 2015a. *Millennium Development Goals and beyond 2015* [Online]. United Nations. Available: www.un.org/millenniumgoals/bkgd.shtml [Accessed 21 December 2017].

United Nations 2015b. The millennium development goals report. New York.

United Nations 2015c. *Sustainable Development Goals* [Online]. United Nations. Available: www.un.org/sustainabledevelopment/ [Accessed 1 February 2018].

United Nations 2017a. *Sustainable Development Goals 4: Quality Education* [Online]. United Nations. Available: www.un.org/sustainabledevelopment/education/ [Accessed 15 March 2018].

United Nations 2017b. *Sustainable Development Goals 7: Affordable and clean energy* [Online]. Available: www.un.org/sustainabledevelopment/energy/ [Accessed 27 February 2018].

United Nations 2018. *Sustainable Development Goals 6: Clean water and sanitation* [Online]. United Nations. Available: www.un.org/sustainabledevelopment/water-and-sanitation/ [Accessed 15 March 2018].

Upton, D. 1983. The power of things: Recent studies in American vernacular architecture. *American Quarterly*, 35, 262–279.

USDA 2018. *Cover crops* [Online]. Available: https://plants.usda.gov/about_cover_crops.html [Accessed 15 February 2018].

Vazir, S., Engle, P., Balakrishna, N., Griffiths, P. L., Johnson, S. L., Creed-Kanashiro, H. et al. 2013. Cluster-randomized trial on complementary and responsive feeding education to caregivers found improved dietary intake, growth and development among rural Indian toddlers. *Maternal and Child Nutrition*, 9, 99–117.

Vázquez, M. A. & Soto, M. 2017. The efficiency of home composting programmes and compost quality. *Waste Management*, 64, 39–50.

Vellinga, M. 2006. The inventiveness of tradition: Vernacular architecture and the future. *Perspectives in Vernacular Architecture*, 13, 115–128.

Viana, C., Coudel, E., Barlow, J., Ferreira, J., Gardner, T. & Perry, L. 2016. How does hybrid governance emerge? Role of the elite in building a green municipality in the eastern Brazilian Amazon. *Environmental Policy and Governance*, 26, 337–350.

Viswanadham, N. & Vedula, S. 2014. Design of Smart Villages. Computer Science and Automation, Indian Institute of Science, Bangalore.

Vyas, D., Hollis, M., Abraham, R., Rustagi, N., Chandra, S., Malhotra, A., Rajpurohit, V., Purohit, H. & Pal, R. 2016. Prehospital care training in a rapidly developing economy: A multi-institutional study. *Journal of Surgical Research*, 203, 22–27.

Wallbott, L. & Rosendal, G. K. 2018. Safeguards, standards and the science-policy interfaces of REDD+: Greening land use through forest-based mitigation in Costa Rica? *Journal of Environment and Development*, 27, 99–125.

Wang, H., Wang, T., Zhang, B., Li, F., Toure, B., Omasa, I. B., Chiramba, T., Abdel-Monem, M. & Pradhan, M. 2014. Water and wastewater treatment in Africa – current practices and challenges. *Clean – Soil, Air, Water*, 42, 1029 – 1035.

Ward, D. 2015. Water requirements of livestock. Ontario: Ontario Ministry of Agriculture.

Waroux, Y. L. P. D. & Lambin, E. F. 2013. Niche commodities and rural poverty alleviation: Contextualizing the contribution of Argan Oil to rural livelihoods in Morocco. *Annals of the Association of American Geographers*, 103, 589–607.

WASHplus 2010. *Zimbabwe – Chiungwiza's "bucket system"* [Online]. Sanitation Updates. Available: https://sanitationupdates.wordpress.com/tag/bucket-latrines/ [Accessed 27 March 2018].

Watson, S. 2018. *Gas vs. electric stoves: Which is really more efficient?* [Online]. Available: https://home.howstuffworks.com/gas-vs-electric-stoves2.htm [Accessed 20 July 2018].

Web of Science 2016. Available: https://apps.webofknowledge.com [Accessed 5 October 2018].

Wei, M., Greenblatt, J. B., Donovan, S. M., Nelson, J. H., Mileva, A., Johnston, J. & Kammen, D. M. 2013. California's Carbon Challenge Phase II Volume I: Non-electricity sectors and overall scenario results. *Scenarios for meeting California's 2050 climate goals.* Berkeley, California: California Energy Commission.

Wells, P. 2007. Deaths and injuries from car accidents: An intractable problem? *Journal of Cleaner Production*, 15, 1116–1121.

Wells, T. N. C., Huijsduijnen, R. H. V. & Voorhis, W. C. V. 2015. Malaria medicines: Glass half full? *Nature Reviews*, 14, 424–442.

White, R. E. 1997. *Principles and practise of soil science.* Carlton: Blackwell Science Ltd.

WHO 1986. Standard model designs for rural water supplies. Copenhagen: World Health Organization.

WHO 2016. Burning opportunity: Clean household energy for health, sustainable development, and wellbeing of women and children. Geneva: World Health Organization.

Wiemann, M. 2013. *Small wind in a developing world* [Online]. Elsevier. Available: www.renewableenergyfocus.com/view/32333/small-wind-in-a-developing-world/ [Accessed 27 February 2018].

Williams, N. J., Jaramillo, P. & Taneja, J. 2018. An investment risk assessment of microgrid utilities for rural electrification using the stochastic techno-economic microgrid model: A case study in Rwanda. *Energy for Sustainable Development*, 42, 87–96.

Wilson, E. O. 2016. Half-earth: Our planet's fight for life, New York: Liveright.

Winter, L., Lehmann, A., Finogenova, N. & Finkbeiner, M. 2017. Including biodiversity in life cycle assessment – State of the art, gaps and research needs. *Environmental Impact Assessment Review*, 67, 88–100.

Wirfs-Brock, J. 2015. *Lost in transmission: How much electricity disappears between a power plant and your plug?* [Online]. Inside Energy Available: http://insideenergy.org/2015/11/06/lost-in-transmission-how-much-electricity-disappears-between-a-power-plant-and-your-plug/ [Accessed 20 February 2018].

World Commission on Environment and Development 1987. *Our Common Future.* Oxford: Oxford University Press.

World Energy Council 2018. *Hydropower* [Online]. London, UK. Available: www.worldenergy.org/data/resources/resource/hydropower/ [Accessed 20 July 2018].

World Health Organization 2015. WHO estimates of the global burden of foodborne diseases. Geneva, Switzerland: Foodborne disease burden epidemiology reference group 2007–2015.

World Health Organization 2018a. *Disease burden and mortality estimates* [Online]. Available: www.who.int/healthinfo/global_burden_disease/estimates/en/index1.html [Accessed 4 April 2018].

World Health Organization 2018b. *HIV/AIDS* [Online]. Available: www.who.int/mediacentre/factsheets/fs360/en/ [Accessed 4 April 2018].

World Health Organization 2018c. *Malaria* [Online]. Available: www.who.int/mediacentre/factsheets/fs094/en/ [Accessed 4 April 2018].

World Wildlife Fund 2018. *Bengal tiger* [Online]. Washington DC. Available: www.worldwildlife.org/species/bengal-tiger [Accessed 22 July 2018].

Wry, T. & York, J. G. 2017. An identity-based approach to social enterprise. *Academy of Management Review*, 42, 437–460.

Yannis, G., Antoniou, C., Papadimitriou, E. & Katsochis, D. 2011. When may road fatalities start to decrease? *Journal of Safety Research*, 42, 17–25.

Yergeau, M.-E., Boccanfuso, D. & Goyette, J. 2017. Linking conservation and welfare: A theoretical model with application to Nepal. *Journal of Environmental Economics and Management*, 85, 95–109.

Yıldırım, M. & Turan, G. 2012. Sustainable development in historic areas: Adaptive re-use challenges in traditional houses in Sanliurfa, Turkey. *Habitat International*, 36, 493–503.

Zhai, Z. J. & Previtali, J. M. 2010. Ancient vernacular architecture: Characteristics categorization and energy performance evaluation. *Energy and Buildings*, 42, 357–365.

Index

For Product Safety Concerns and Information please contact our EU representative GPSR@taylorandfrancis.com
Taylor & Francis Verlag GmbH, Kaufingerstraße 24, 80331 München, Germany

www.ingramcontent.com/pod-product-compliance
Ingram Content Group UK Ltd.
Pitfield, Milton Keynes, MK11 3LW, UK
UKHW021302190926
475701UK00009B/582
9780815365655